Viktor Kuzmin
Lyudmila Ognichenko

MOLECULAR INFORMATION FIELD

Viktor Kuzmin
Lyudmila Ognichenko

MOLECULAR INFORMATION FIELD

FUNDAMENTAL ASPECTS AND APPLICATIONS

ScienciaScripts

Imprint
Any brand names and product names mentioned in this book are subject to trademark, brand or patent protection and are trademarks or registered trademarks of their respective holders. The use of brand names, product names, common names, trade names, product descriptions etc. even without a particular marking in this work is in no way to be construed to mean that such names may be regarded as unrestricted in respect of trademark and brand protection legislation and could thus be used by anyone.

Cover image: www.ingimage.com

This book is a translation from the original published under ISBN 978-620-6-15502-7.

Publisher:
Sciencia Scripts
is a trademark of
Dodo Books Indian Ocean Ltd. and OmniScriptum S.R.L publishing group

120 High Road, East Finchley, London, N2 9ED, United Kingdom
Str. Armeneasca 28/1, office 1, Chisinau MD-2012, Republic of Moldova, Europe
Printed at: see last page
ISBN: 978-620-8-27823-6

Contents

INTRODUCTION

With the development of human society, the role of information in human life has significantly increased, and with the advent of computers, the volume of processed information has significantly increased. Information confidently penetrates into all spheres of human activity, in this regard there is a need for quantitative assessment of information contained in the object of research.

In organic chemistry, the construction of various information indices quantitatively characterising the molecular structure is of great importance, which plays an essential role in the encoding of structural information and allows optimal formation of databases. At present, one of the most important tasks of modern theoretical organic chemistry is also the study of the problem of quantitative structure-property relationship of organic compounds (QSAR/QSPR - Quantitative Structure - Activity/ Property Relationship). The solution of such a problem is of undoubted interest, since it allows the purposeful synthesis of compounds possessing with a high degree of probability the given properties.

The use of computer technologies and information characterising research objects opens up the possibility to carry out not only preliminary screening, but also molecular design of organic substances exhibiting one or another property on the basis of the revealed regularities. All this makes it possible to significantly reduce the enormous expenditure of time and resources required for the development of new substances with a set of specified properties.

Thus, the development of methods that allow generating new types of parameters reflecting information about the investigated objects and using information characteristics as structural descriptors to solve various "structure-property" problems is currently topical.

The authors of the monograph have developed an original concept of the information field (IF) of a molecule, which opens a new view of the molecule as an object forming information in the surrounding space. In this regard, the concept of IP goes beyond chemistry and can be applied to any object (physical body). However, the authors are professional organic chemists and therefore focused their attention on fundamental aspects of molecular structure and its influence on the properties of organic compounds, using the possibilities of the developed IP concept.

It should be especially emphasised that the proposed concept does not use any esoteric theories, supernatural phenomena and the like. Our concept of IP is based only on natural-scientific ideas, which are based on Shannon's probabilistic approach to measuring the amount of information.

The main objective of our book is to show the possibilities of the developed FE concept for solving a variety of chemical problems. Given the universality of our concept, we hope that specialists of other fields will be able to transform our ideas for their problems where other (non-molecular) objects are considered.

CHAPTER 1

MOLECULAR STRUCTURE IN TERMS OF INFORMATION THEORY

1.1 Information theory in organic chemistry

The term "information" has been known to mankind since ancient times. In Latin, it means "explanation", "statement".

As mankind has developed, the need for people to exchange information with each other, as well as the need to preserve the available information, has continuously increased. This need was realised with the invention of writing, radio, telephone, television and other ways of transmitting information, and the creation of various devices that record or store information. As a consequence, a whole science has emerged, studying information, ways of its transmission, storage and processing.

However, what is information? It is not very easy to answer this question at present, and the definitions of this concept available in the scientific literature are very diverse and contradictory. Some researchers consider information as a set of information about certain phenomena [1-6], and others consider information as a quantitative measure of orderliness of structures, i.e. as a measure of system organisation [7-13]. Any known natural phenomena corresponds to human information about these phenomena. Although information always "belongs" to a particular object-carrier, however, it does not directly follow from its nature (is not determined by its essence), i.e. the quality of information is relatively independent of the nature of the object-carrier of this information. It is this relativity that leads to the fact that the same information can be transmitted by telephone or television, recorded on a photograph, written on paper, carved on a rock or stored in memory. Thus, all natural phenomena can exist either directly, i.e. in their inherent material form, or indirectly, through special carrier objects. Since information about a given object has a different material expression than the phenomenon itself, it cannot and does not possess all the properties of this phenomenon, is not identical, but only corresponds to it. We can say that information is the existence of various phenomena in the form of displays or messages [14].

Information can be considered as the essence of life, the essence of species diversity of natural forms, the essence of the quality of human abilities [15]. Information can preserve and destroy health, form natural cataclysmic processes, military conflicts, ecological disasters. Information builds life in its beingness and eventfulness, in its tension and criticality. In this regard, there is a need to protect and manage information in order to solve the problems of humanity, including those lying in the sphere of health, ecology, creative self-development.

The emergence of modern information theory dates back to 1948. It was in this year that K. Shannon's famous work "Mathematical Theory of Communication" was published, in which a formula is given to determine the amount of information contained in any message [16]. Information and uncertainty are terms describing any process that can be represented as a set of objects. If there is some uncertainty, in order to get information we need to remove this uncertainty and the more uncertainty, the more information is needed to remove it. Information is a reduction in the uncertainty of the receiver (or molecular

machine). Thus, we can measure the amount of information through the measurement of uncertainty, which is a constant characteristic of the probability of the state of systems.
With any spontaneous transformation in nature, disorder tends to increase, entropy increases. We need information to reduce entropy and hence, the greater the entropy, the more information is required for "orienting" [17]. It is known that what is homogeneous, lacking diversity, cannot contain information. Information science studies what can be realised due to the degree of diversity in any system or energy flow [18].
Definitions of the basic concepts of information theory are given by Schneider in the Glossary of Information Theory [19].

1.1.1 Application of information theory to the characterisation of structural diversity of molecules.

The world of molecules is characterised by great diversity. Molecules of organic compounds are characterised by three main parameters: composition, structure and properties. One of the key problems in chemistry, physics and other fields of science is the problem of the relationship between composition, structure and properties, which is closely related to the question of chaos and order and the relationship between them. The triad of composition - structure - properties should be supplemented by an integral numerical (scalar) characteristic of a molecule, which reflects its structure, on the one hand, and is simultaneously a characteristic of chaos and order and their correlation, on the other hand. In this case it becomes possible not only to describe individual molecules in the most complete way, but also to carry out their general systematisation and comparison in different series with the help of modern information theory and molecular graph theory. Science endeavours to represent these parameters in the form of some quantitative certainties. A quite rigorous, numerical measure may be found for the elementary or isotopic composition of a molecule. Each of its properties - whether it be heat content or ionisation potential, basicity or polarizability, dipole moment or colour - is described by some quantity. As for the structure, it has to be stated that there is no quantitative, numerical measure characterising the structural appearance of a molecule.
For the chemist, the structure of a molecule is a geometrical model, a spatial construction, although possessing within itself some measurable indices, such as interatomic distances or bond angles, but lacking an integral quantitative characterisation.
Thus, the problem is reduced to trying to find some abstract value that would describe the molecular structure in an integral way. Such a value, apparently, should fix the peculiarities of the structure, the differences of one molecular structure from another, their diversity and variety; it should be a measure of a certain order inherent in a given multi-atomic aggregate.
In natural science, entropy is a familiar measure of the inhomogeneities and differences existing in a system, a measure of its unevenness and diversity. Without in any way rejecting the possibilities contained in thermodynamic entropy as a characteristic of molecular structure, it is possible to use another measure of orderliness, which was proposed by K. Shannon and defined by him as the entropy of information [16]. Shannon and defined by him as information entropy [16].
In 1928, Hartley, an American engineer, set out to quantify or measure the diversity of a set of events. Translating Hartley's reasoning into chemical language, let us ask ourselves,

what is the measure of diversity of composition of a typical amino acid molecule consisting of carbon, hydrogen, oxygen and nitrogen atoms? Obviously, the atoms in an amino acid come in four varieties. In other words, if we conduct an experiment to determine the element of the qualitative composition of an amino acid, this experiment has 4 possible outcomes.

Suppose now that we set up an experiment in which we can detect only one of the elements that make up an amino acid. In such a case we are faced with an uncertain situation before starting the experiment, since the element detected could be any one of the four. To determine the degree of uncertainty of an experiment with k different outcomes, Hartley proposed the value *log k* as a quantitative measure [20]. Then for a typical amino acid the measure of qualitative diversity is $log4 = 0.6021$.

It should be noted that this approach is limited because it cannot be used to distinguish one amino acid from another in any way. Hartley's approach makes no distinction between the quantification of different outcomes: a chemist would say that it ignores data from quantitative analyses of different elements.

It will be possible to take into account quantitative differences between individual outcomes of experience if we follow Shannon and introduce the entropy of information established by him as a measure of diversity. Let us consider an experience with possible outcomes $A1, A_2 \ldots \ldots, A_k$ Let $p(A1), p(A_2), \ldots, p(A_k)$ *are the* probabilities of the individual outcomes of the experience. According to Shannon, an uncertainty equal to - $log\ p(Ai)$ should be assigned to each outcome *A 1 of* experience *a.* For all possible outcomes, the value $H(a)$, *which* is the sum of uncertainties of all possible outcomes, is introduced as a measure of uncertainty.

$$H(\alpha) = -p(A_1)\log p(A_1) - p(A_2)\log p(A_2) - \ldots p(A_k)\log p(A_k) =$$

$$= -\sum_{i}^{k} p(A_i)\log p(A_i).$$

Obviously, when all $p(A_t)$ are equal, Shannon's information entropy reduces to Hartley's formula.

Let us now apply Shannon's formula to the amino acids. Let us compare, for example, two of them: glycine (C2H5O2N) and glutamine (C5H1OO3N2). The outcome of the experiment is now determined by the probability of finding one or the other atom among all the atoms that make up the molecule. For glycine:

$$H_{глицин} = -(2/10)\log(2/10) - (5/10)\log(5/10) - (2/10)\log(2/10) -$$

$$- (1/10)\log(1/10) = 0{,}5611.$$

For glutamine:

$$H_{глутамин} = -(5/20)\log(5/20) - (10/20)\log(10/20) - (3/20)\log(3/20) -$$

$$- (2/20)\log(2/20) = 0{,}5242.$$

The results obtained show that the information entropy of these molecules is different. It should also be noted that it is lower than the Hartley entropy value.

The question arises why the measure introduced by Shannon is called information entropy. This problem is considered in detail in [17, 21]. We will touch upon it by analysing the behaviour of the function *H.* Let us imagine that only one of all possible outcomes of the experience is realised. Then for it $p\ logp = log\ 1 = 0$. But all other summands also turn to zero, since for all of them $p = 0$. As a result, we obtain $H = 0$, which means the absence of any uncertainty of experience.

If all outcomes are equally likely, i.e. all $pi = p$, then the value of H reaches its maximum. Figure 1.1 shows the Shannon entropy curve. When all atoms are the same, say X, the entropy is 0. If the atoms change (i.e., are reassigned) from X to Y, the entropy increases until it reaches a maximum value when the number of atoms Xi Y is the same. The entropy value then decreases as the remaining X atoms are changed, "converting" to Y atoms until there are no X atoms left at all and all atoms are labelled as Y.

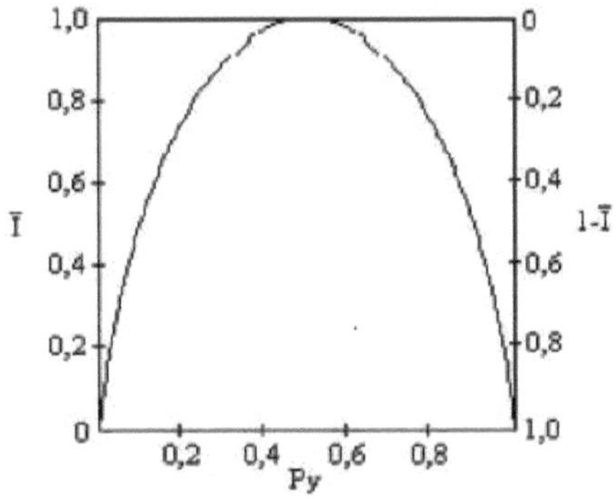

Fig.1.1 Shannon entropy curve.

The physical entropy S behaves in a similar way: it is minimal at maximum non-alignment of the system and tends to the maximum when any differences and inhomogeneities in the system disappear. It also draws attention to the fact that the analytical expression for H *is* similar to Boltzmann's definition of entropy. Boltzmann associated entropy with the amount of disorder in the system and with the "loss of information". Shannon generalised this concept to include all types of information, and Movšović related it to the complexity of graphs [22]. The relationship between complexity and disorder implies that the more disordered a system is, the more complex the problem of its description (i.e., more information is needed).

The Shannon function describes the uncertainty of the outcome of our experience and thus determines the amount of information that we eventually receive. P. Chambadal noted that, in general, the value of H denotes the lack of information before solving a problem or, what is the same, the information we receive after solving the task [23].

The connection between entropy and information is revealed in his works by A.N. Kolmogorov [24, 25]. Let a variable be able to take values belonging to a finite set X, which consists of N elements. Then the entropy of the variable x is equal to:

$$H(x) = log2N.$$

By specifying a particular value of $x = a$ *to* the variable x, we remove this entropy by communicating information:

$$I = log2N.$$

If the variables $x1$, x_2 , ... x_K are capable of independently traversing sets that consist of

$N1, N_2, \dots$, respectively. , N_K elements, then

$$H(x_1, x_2, \dots x_K) = H(x_1) + H(x_2) + \dots + H(x_K).$$

A. N. Kolmogorov's conclusion that the basic concepts of information theory can be justified without recourse to probability theory and in such a way that the concepts "entropy" and "quantity of information" turn out to be applicable to individual objects [25] is extremely important for us.

Another important condition for applying information theory to research objects is the way of distinguishing these objects. W. Ashby [26] drew attention to this circumstance: "Difference is the objective basis of information". Undoubtedly, when determining the entropy of information, it is necessary to specifically stipulate what features, criteria of distinction are taken into account.

The necessity of strictly distinguishing the feature, on the basis of which the feature of an object is fixed, led to the formulation of the signature principle in biology [18]. This principle, formulated by G. Kastler, allows measuring information by selecting in an object only a part of its inherent features, a sign, a mark. In this case, Kastler proceeds from the fact that in most real cases it is necessary to take into account only a part of the inherent features of the object in order to transfer the information contained in a biological object [18]. V.M. Glushkov in his work [27] also expressed the idea that any heterogeneity carries some information and at the same time this heterogeneity should be strictly fixed in its qualitative form.

The above mentioned allows us to approach the solution of specific chemical problems related to the structure of a molecule.

The structure of a molecule can be simplified by considering it from the perspective of graph theory [28, 29]. A graph is a representation of some set of points for which the way of their connection is specified. A graph $G = (V,E)$ consists of a finite set V, whose elements are called vertices, and a set E of two-element subsets of V; the elements of the set E are called edges.

Molecular structures are actually graphs in which atoms are represented by vertices and covalent chemical bonds by edges. The number of edges whose end is a given vertex is called its degree. Two vertices can be connected by several edges, and then such edges are called multiples. Such a graph describes the connectivity of atoms in the molecular skeleton independently of the metric properties, i.e., equilibrium internuclear distance, valence angles, etc., corresponding to a given chemical structure. Consequently, the graph-theoretic description reflects those features of the molecular structure that depend on connectivity, as opposed to properties due to the precise geometric arrangement of the constituent atoms of the molecule in space. It is in this sense that chemical graphs are topological (rather than geometric) representations of molecular structures [30]. Molecular graphs are graphs of a special kind, and the main problem is to assign a code to each vertex such that two graphs $G1$ and $G2$ have the same code if and only if $G1 = G2$.

Let us consider the process of transition from chemical structure to molecular graph on the example of pyran:

At this transition hydrogen atoms turned into first degree vertices; oxygen atom - into a vertex of degree 2 and carbon atoms - into vertices of degree 4. Simple chemical st - bonds were transformed into simple edges, and double bonds appeared in the form of multiple edges.

A graph that can be represented in the plane in such a way that it does not contain intersections of edges is called a planar graph. Chemical graphs are usually categorised as planar graphs. A sequence of edges following each other is called a chain. If any two vertices of a graph are connected by a chain, then such a graph is called a connected graph. Using the proposed definitions, we can describe many chemical structures in the form of connected planar graphs.

Our next step will be to introduce some general quantitative characteristic for molecular graphs. Obviously, it should be a measure of the diversity of vertices of a given graph. This idea was expressed by N. Rashevsky in 1955, developed by E. Tracco and the Valentinuzzi brothers [31] The essence of it is as follows.

Each graph contains some finite amount of information. In the information theory approach, the corresponding set A of n elements, obtained from a molecular graph, is partitioned into h non-overlapping subsets A_t of power n_i $(i=1,2,...,h;\ \sum_i n_i = n)$ using an equivalence relation defined on the set A. Then a probabilistic scheme can be related to this distribution

$$\begin{pmatrix} A_1, A_2, \dots A_h \\ p_1, p_2, \dots, p_h \end{pmatrix}; \quad \sum_i p_i = 1; \qquad p_i \geq 0; \quad (i=1,2,\dots,h);$$

$$p_i = \frac{|A_i|}{|A|} = {}^{n_i}\!/_{n} \quad (i=1,2,\dots,h);$$

where *pi is the* probability that a randomly selected element of set A will be in the *i-th* subset.

The entropy (or complexity) of the structure associated with this type of partitioning is calculated using Shannon's formula:

$$H = -p_i \log_2 p_i \text{ bit.}$$

Consider how the entropy of information is determined for individual chemical graphs.

Thus, the graph of the benzene molecule contains 12 vertices, of which six identical vertices for carbon atoms and six for hydrogen. In this case, the probability that a given vertex is a carbon atom is . $p_c = \frac{6}{12} = \frac{1}{2}$.

The same value is the probability $p_н$ of meeting a hydrogen atom in the vertex.

Count for benzene Count for methane

For a methane molecule: $p_c = 2\ ar_н = 4\ 5^5$

If the vertices of the graph are labelled, i.e., if we take into account the diversity of atoms, it is possible to determine the entropic characteristics of more complex molecules. Thus, for chloromethane: $p_c = |;\ p_н =^3;\ p_d = .^1$

For each of the above graphs, it is easy to calculate the entropy of information.

$$H_{бензол} = -\frac{1}{2}log_2\frac{1}{2} - \frac{1}{2}log_2\frac{1}{2} = 1$$

$$H_{метан} = -\frac{1}{5}log_2\frac{1}{5} - \frac{4}{5}log_2\frac{4}{5} = 1{,}58$$

$$H_{хлорметан} = -\frac{2}{5}log_2\frac{1}{5} - \frac{3}{5}log_2\frac{3}{5} = 1{,}43$$

Along with entropy for chemical graphs we can use the term "information content" borrowed from the theory of message transmission.

Based on the values of the information content of molecular graphs, it is possible to classify different structures and determine the change in the information content during chemical transformations. Thus, at condensation of three molecules of acetylene

$$3\ C_2H_2 \rightarrow C_6H_6$$

$$\Delta H = H_{C6H6} - 3H_{C2H2} = -2\ бита$$ bat

It is interesting to note that a decrease in thermodynamic entropy is also observed for this reaction:

$$\Delta S^0 = 64 - 3{\cdot}48 = -80\ эв.$$ ev.

As we have already noted, not only the degree of a vertex but also its position in the graph characterise the measure of its diversity. Let us compile the information content of cyclooctatetraene and bicyclo-octatriene related to it by reversible transition [32]

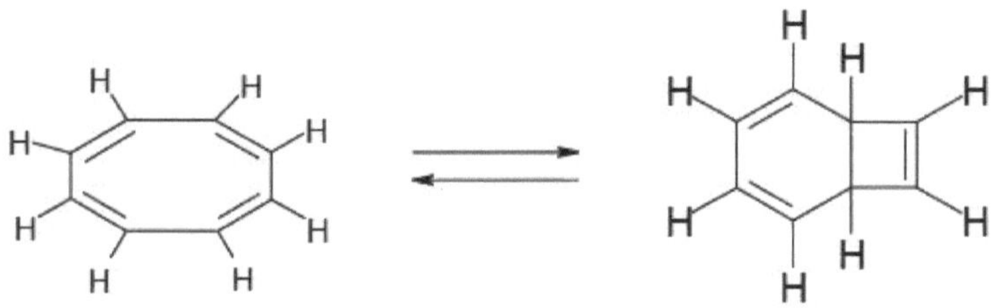

While in cyclooctatriene all carbon atoms are identical, in its bicyclic isomer they form four pairs differing in position in the molecule. In this case:

$$H = -4 \cdot \frac{2}{16} \log_2 \frac{2}{16} - \frac{8}{16} \log_2 \frac{8}{16} = 2$$

Since $H=1$ for the original cyclooctatriene, we can say that the diversity measure for the bicyclic isomer is higher, the uncertainty of the outcome of our experience has also increased if we randomly select a single atom.

The disadvantage of representing molecules using graphs is that all stereochemical features of the molecular structure are lost. However, graphs describe the complete topology of a molecule. It is known that many important characteristics of molecules, such as, for example, energy, charge density, and bonding order depend significantly on topology [22].

Using the described approach, it is possible to determine the information content of different isomeric compounds depending on the degree of branching of carbon chains.

In biochemistry and organic polymer chemistry, we have to deal with branched molecules composed of identical subunits. In order to assess the specificity of action of enzymes or catalysts that synthesise or cleave them, it is useful to determine a quantitative measure of the branching of the molecule.

Let us consider initially on an elementary example how this characteristic is determined. Suppose that a molecule has only one branching node of type *A:*

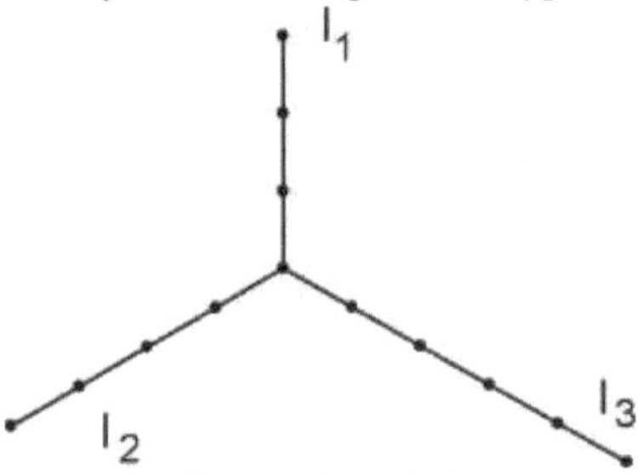

Then we will determine the degree of branching of this node based on the relative lengths

of segments *11*, l_2, l_3. In the considered case $P_i = \frac{l_i}{\sum l_i}$ and

$$H = -\frac{4}{12}\log_2\frac{4}{12} - \frac{3}{12}\log_2\frac{3}{12} - \frac{5}{12}\log_2\frac{5}{12} = 1{,}544$$

It is quite obvious that the degree of branching will be the same when the length of the side chains changes proportionally. It follows from the general properties of the Shannon equation that the maximum branching will be at the equality $P1=P_2=P_3$. In this case

$$H = -\frac{1}{3}\log_2\frac{1}{3} - \frac{1}{3}\log_2\frac{1}{3} - \frac{1}{3}\log_2\frac{1}{3} = 1{,}585$$

It is interesting to compare the degree of branching of other *A* isomers, e.g.

B1

B2

For *B1* we have $H=0.815$, for B_2 we have $H = 1.324$. The intuition of the chemist also testifies in favour of such a ratio of the degree of branching in the depicted isomers.
The degree of branching obviously increases if the node will have not three but four segments. Indeed, in the case of isomer *A of the* following structure

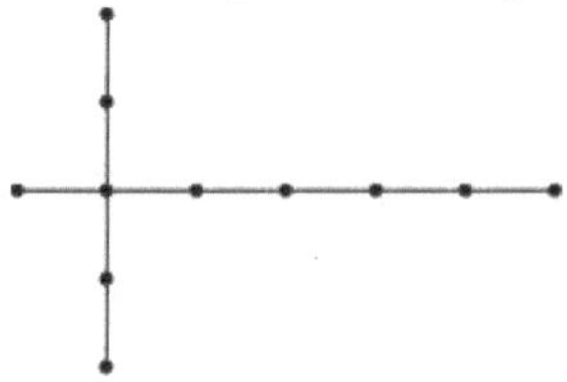

we have $H = 1.782$.
A much more complicated situation occurs when there are several branching nodes in a molecule. In such a case, we will take the sum of the degrees of branching of each node as a measure of the branching of the molecule, and for all nodes *k* we have $H=\sum H_k$. In determining H_κ of each branching node, we will consider only the number of structural units in the segment.
For example, for a molecule of the following structure

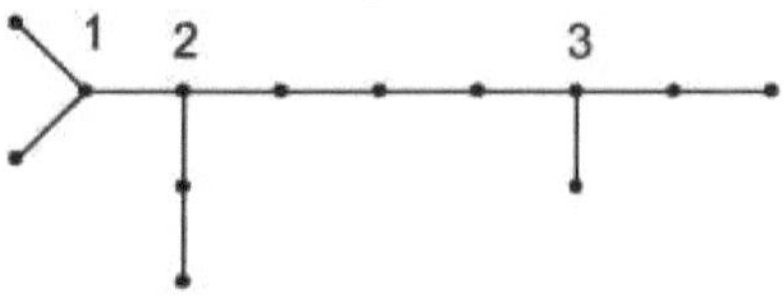

calculated degree of branching taking into account three nodes $H= 3.247$.
That is, we have considered a quantitative characterisation of the structure of a molecule, abstracting from its specific structure, reducing it to a set of vertices and edges of molecular graphs.
Thus, molecular structure has been a very fertile field of application of information theory. Molecules are easily formalised in the form of various discrete models, for each of which it is easy to calculate the information content (Fig. 1.2) [33]. It is clear that depending on the model within which a given molecular structure is represented, the information content may be different. As can be seen from Fig. 1.2, the simplest model of molecules, the molecular graph (a), contains a minimum of information. Taking into account the nature of atoms in the molecular structure (structural formula) naturally leads to an increase in the amount of information (b). And differentiation of atoms of the same nature depending on their neighbours also contributes to the growth of the amount of

information (c). Spatial models describe the molecular structure most adequately and their information content is maximum (d, e).

Model	Groups of equivalent elements of the structure	Amount of information (in bits)
а) 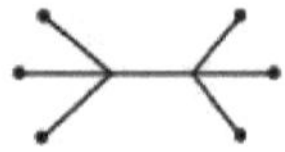	(2,6)	0.81
б) C_2F_4ClH	(4, 2, 1, 1)	1.75
в) 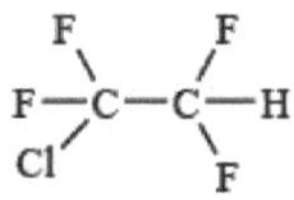	(2, 2, 1, 1, 1, 1)	2.50
г) 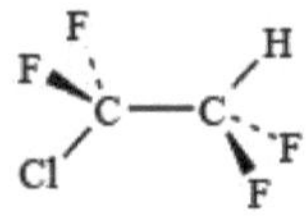	(2, 2, 1, 1, 1, 1)	2.50
д) 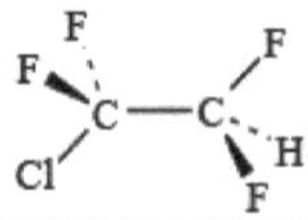	(1, 1, 1, 1, 1, 1, 1, 1)	3.00

Fig. 1.2 Information content of the molecular structure

1.1.2. Application of information theory to the characterisation of dynamic states of molecules.

Using information theory it is also possible to analyse the internal dynamics of a molecule. Thus, Mazur notes that as a message, information theory can consider a physical state that differs from other physical states in a certain way [34]. He specifies this position, drawing attention to the fact that any physical state can be considered not only from the statistical point of view, describing the distribution of matter in space, but also from the energy-material point of view, taking into account the distribution of energy and matter in space and time.

Similar views were held by Norbert Wiener: "With the passage of time, any state of the system can and will change into any other state. However, the probability of such a transition depends ultimately on the relative probability or measure of these two states" [35]. [35].

As a generalised characteristic of the dynamic structure of an organic molecule, it is necessary to introduce such a parameter that would reflect the structural uniqueness of the system, and thus its reactivity, in a generalised form accessible to quantitative evaluation.

A molecule of an organic compound can be considered as a carrier of some information related to its structure. In this case, the question arises about the estimation of the structural information contained in the molecule. In the most general form, such a quantitative measure would be the total information capacity of the molecule, which is determined by the number of discrete states, i.e., the number of all possible isomers composed of given atoms. Such a definition is of little use because a huge mass of isomers is not realised, so for a realistic assessment of the information in a molecule one should select such discrete states that the molecule adopts under ordinary conditions and that are essential for its chemical behaviour.

Such information is called active information. This value turns out to be a homogeneous measure for chemical structures and opens the possibility of their comparison and relative evaluations. In order to determine the information content of the dynamic structure of a molecule (dynamic information capacity), it is necessary to determine the number of possible states of the molecule and their statistical weight. Rakov [31], on the basis of group theory, estimated the information capacity of the asymmetric carbon atom to be 1 bit per molecule, or 1 molbit. I.e., using the Shannon equation, for an equal probability realisation of D- and L- configurations we have:

$$H = -\sum_i p_i \log_2 p_i = -0{,}5\log_2 0{,}5 - 0{,}5\log_2 0{,}5 = 1$$

Since the total amount of information contained in the whole, for independent parameters, is equal to the sum of the amounts of information contained in its parts, the active information capacity of an organic molecule can be defined as the sum of information entropies for each discrete state:

$$Q_a = \sum_j H_j$$

The following discrete states of molecules are chosen as active, in the informational sense: atomic asymmetry, molecular dissymmetry, cis-trans isomerism, stable conformations, and tautomerism.

Let us consider the method of determining Q_a for amino acids. In the case of asparagic acid, possible conformations of carboxyl groups and internal salt formation by proton transfer to the amino group from both carboxyls must be taken into account. Then

$$Q_a = 2H_{OH} + 2H_{C=O} + H_{C_2-C_3} + H_t + H_{as} + H_{NH_3} = 10{,}36.$$

If there is an aromatic or heterocyclic radical in the molecule (in phenylalanine, tryptophan, etc.), its two rotational isomers are taken into account.

It is necessary to establish such a criterion of information capacity, which does not depend on the size of the molecule. Since not only carbon atoms but also other atoms of the molecule take part in the formation of active information capacity, the concept of specific information capacity of compound *IA* per one atom *A is* introduced:

$$I_A = \frac{Q_A}{\sum A_i} = \frac{\sum H_j}{\sum A_i} \text{ bit/atom.}$$

The value of I_A serves as an objective criterion of structural richness of organic compounds. It varies naturally within a given class of organic compounds and is a characteristic of the class as a whole.

1.1.3. Relation of information entropy to symmetry and similarity.

Information entropy is quantitatively related to similarity and symmetry. A greater entropy value correlates with greater molecular similarity. The maximum entropy of any system corresponds to the total loss of information, the highest symmetry and the greatest simplicity.

A logarithmic dependence between entropy and symmetry was established by Lin [36]. In fact, the correlation between entropy and symmetry in qualitative form was hinted at by Schrödinger many years ago: negative entropy corresponds to asymmetry, broken symmetry, or lower symmetry.

Correlations can be represented in the following logarithmic form:

$$S = -\ln \sigma,$$

where for simplicity the positive constant *k* (Boltzmann constant) is equated to 1, st is the symmetry number for an individual molecule.

Arguments based on the fact that a decrease in information content corresponds to an increase in entropy make it possible to show that, in general, the correlation between entropy and symmetry is simply represented by a logarithmic expression:

$$S = \ln w_a,$$

where w_a symmetry number is the number of symmetric or indistinguishable microstates.

In this equation, Lin omitted the minus sign, hence entropy should increase with increasing symmetry number.

The algorithmic concept of entropy can be compatible with the usual informational concept of entropy. If a system evolves towards increasing symmetry, the mathematical expression describing the system will naturally simplify. Consequently, in this case the information content of the system decreases and the algorithmic entropy increases. The established logarithmic entropy and symmetry relations were used to quantitatively interpret dynamic stereochemical phenomena in chemical reactions and phenomena accompanied by symmetry breaking in crystallisation phase transitions. These logarithmic relations can also be used in many other cases, taking into account various phenomena of structural stability and spontaneity of the process, to study any systems of complex hierarchical dynamic and static structures.

Various methods based on the principles of Shannon information theory have been widely used to estimate molecular diversity [37-45]. In particular, one such method was proposed by Lin [41]. In this approach, a series of compounds is considered as a static molecular set or a set of microstates that contain information about their environment with respect to some particular property. *The* molecular diversity is directly related to the information content of the set (*I*), which is defined by the following formula: $I = S_{max} - S$, where *S is*

the entropy of the system calculated using the classical Shannon entropy equation.
In Lin's approach, each composite set represents a finite number of different molecular subsets. The entropy of the system is determined by the equation :

$$S(m,n) = -\sum_{j=1}^{n}\sum_{i=1}^{m} p_{ij} \ln p_{ij}$$

where *m is the* number of subsets,
n is the number of elements in the set,
Pij is the probability of finding the *i-th* element in the *j-th* subset.
The value of *ru* can be calculated using molecular similarity tables [41], it takes values in the interval [0;1].
According to Lin, as the probabilities converge, variety becomes more indistinguishable, the entropy of the system increases and information decreases.
The problem of applying information theory to organic molecules has also been addressed by Graham [42]. Here, a statistical approach covering all possible molecules and states is proposed, in which chemical structures are considered through integer statistical calculations, the development of an integer variable determining the total number of building complexons of an organic molecule is discussed.

1.2. Biological development and information theory

The biological theory of evolution can be translated into the language of information theory. Such work, which gave very instructive results, was carried out by I.I. Shmalhausen [43]. With the help of information theory, Shmalhausen revealed the meaning of direct and inverse relationships that determine the evolutionary process, and at the same time he emphasised that for biology the usual information theory is completely insufficient. Conventional theory considers only the amount of information contained in some message, regardless of its meaning. Thus, we get one bit of information when answering any question posed in the form of "yes" or "no", i.e., when solving a binary alternative, whether it is a question about the result of tossing a coin or a question about the validity of the second beginning of thermodynamics for living systems.
The value of information is determined by the events caused by the reception of the message and can be expressed through the consequences of the reception. The occurrence of these events means a change in the perceiving system. If the system does not change under the action of the message, there is no reception. It is obvious that information reception is an irreversible process occurring in a non-equilibrium system characterised by the presence of unstable states. Receipt of information moves the system to one of the relatively stable states, in equilibrium conditions there is no reception and there is no value of information.
Since receptivity is a multilevel process, the value of a message depends on the level of receptivity: for one receptor the message may have a high value, while for another it may have zero value. At this level, only non-redundant information, i.e. information that is not yet known to the receptor and that it can perceive, has value.
Consider, for example, a language message. If the number N of different letters in a language is known, then the amount of information per letter is $I_1 = \log_2 N$.
For Russian language $N = 32$ and $I = 5$ bits. This is the first level of receptivity. At the

next level, the probability of occurrence of a given letter is taken into account (e.g., the probability of encountering the letter "a" is higher than the letter "i"). Using Shannon's formula, $i = N$

the amount of information per letter $I_2 = -\sum_{i=1}^{i=N} p_i \log_2 p_i$, with the sum of all probabilities equal to one: $\sum_{i=1}^{i=N} p_i = 1$.

It can be seen that $I_2 < I_1$ - the amount of information per letter has decreased, since there is already information about the frequencies of occurrence of letters. At this level, the message contains redundant information. Redundancy is determined by the following formula:

$$R_2 = 1 - \frac{I_2}{I_1}.$$

At the next level, pairwise correlations between letters are taken into account (e.g., in Russian, a vowel more often follows a consonant than a vowel), followed by triple, quadruple and higher correlations.

As the level of receptivity increases, the redundancy of information increases and the value, the indispensability of the remaining non-redundant elements of the message, increases. Based on this, Wolkenstein [44] formulated a conditional definition of the value of information.

Let the message contains $n1$ letters and the amount of information $n1I1$. At the next level, the amount of non-redundant information is reduced to $n1I_2$. We replace the reduction in the amount of information per letter by a reduction in the number of letters in the message to n_2, i.e. $n_1 I_2 = n_2 I_1$. The effective number of letters has decreased by $\frac{I_1}{I_2} = \frac{1}{1 - R_2}$ times, and the information per letter, i.e. its value, has increased by the same amount.

It is necessary to distinguish between the reception of information entering the perceiving system and the creation of new information in the system. According to G. Kastler [18], the creation of new information is the memorisation of a random choice. Kastler gives the following example: a person leaves a suitcase in an automatic storage chamber, types in some numerical cipher and memorises it. It is possible to open the camera only knowing this cipher, i.e. information is created.

In biology, the creation of new information can be considered, for example, the emergence of each new individual as a result of sexual reproduction, nothing pre-programmed, random recombination of genes, which leads to a very wide variability, giving unlimited material for evolution.

A biological system is characterised by the presence of a whole hierarchy of information reception levels. During ontogenetic development, the specialisation of information messages, their irreplaceability and, consequently, the value of information increases. At the same time, the amount of information at higher levels of organisation increases, i.e. along with the initial genetic information, new information is created, encoded not by the sequence of nucleotides, but by the organisation of cellular structures, morphology of the individual, etc.

In connection with the evaluation of information by its irreplaceability, it is possible to determine the value of a message element by the consequences of replacing that element. The more significant are the changes introduced into the receptor system by the replacement of an element, the higher is its value. In the course of biological development of an organism as a whole, the amount of information increases. The development goes from simple to complex (complexity can be defined by the number of words needed to describe a biological system). The ability of the system to select valuable information also increases. The loss of adaptive capabilities associated with the simplicity of structure and interchangeability of cells and tissues is more than offset by the vital advantages of an autonomous complex system. Autonomy arises precisely as a result of the selection of valuable information. The ability to select reaches the highest level in higher animals, whose sense organs are designed for this purpose. A frog reacts only to a moving insect, a bat or a dolphin, using ultrasonic echolocation, reacts only to reflected signals.

As we have already mentioned, one bit of information is equivalent to a change in entropy by *kln2*. this means that to obtain one bit of information it is necessary to expend energy not less than *kTln2*. However, *kTln2* is spent on one bit of any information, both valuable and non-valuable, redundant and substitutable. Selection of valuable information does not require additional energy expenditures. For this purpose it is sufficient, for example, to arrange the channels in the membrane in such a way that molecules or ions of a certain shape and size pass through them. Energy costs were incurred at the previous stages of biological evolution. That is, the main features of a living system in the aspect of information theory are the ability to create new information and select valuable information without special energy costs.

1.3. Use of various information indices that quantitatively characterise the structure of molecules

Rapid computerisation of chemistry and chemical technology, efficient storage and use of chemical information in computers makes the issue of coding this information extremely relevant [45]. At present, a number of indices quantitatively characterising the structure of molecules and allowing optimal formation of databases have been proposed. Such indices represent various functions of the total number and type of atoms, groupings, rings, and geometrical features of molecules, and most of the coding is based on methods of information theory and graph theory.

Currently, there are many methods available for obtaining numerical indices of molecules, which correlate quite well with the most important physicochemical characteristics of compounds and can be used for quantitative description of molecules in the study of their physicochemical and biological properties.

As already mentioned, information theory allows us to calculate a measure of information - the entropy of information in Shannon's phrase:

$$H(x_i) = -\sum_{i=1}^{N} p_i(x_i)\log_2 p_i(x_i), \sum p_i = 1, \qquad (1.1)$$

where *H is the* entropy of information, p_i is the probability of the outcome of experience *xi,* which is calculated in each case.

If we take into account, for example, only the chemical composition of a molecule, the calculation according to the above formula gives the corresponding information index of chemical composition. Using only the topology of the molecule as represented by its structural graph, we obtain an information and topological index. Many important characteristics of molecules, such as energy, bond order and charge density, depend significantly on topology and since topological indices are numerical expressions of certain topological properties of the molecular structure, it is not surprising that different topological indices correlate significantly with the physicochemical and biological properties of diverse groups of molecules [22].

Molecular negentropy is used to encode structures controlling biological activity. A preparative molecule is considered as an information source with information content available to the receiver. This index evaluates molecules according to symmetry and encodes structural characteristics that affect physical properties and biological activity in certain cases. As a measure of the information content of a molecule, molecular negentropy has been successfully used as a structural parameter to quantitatively analyse molecules with some uncertain biological response [46, 47]. For example, [46] provides quantitative correlations between the negentropy of various molecules and their: a) toxic effects on Madison 517 fungus; b) narcotic effect on arenicola larvae; c) destructive effect on turtle heart; d) narcotic effect on frog heart.

Shannon entropy (SE) and Shannon entropy differential (DSE) are used to analyse different molecular descriptors in large compound databases [48]. This allows the detection of inherent differences between descriptors in compound databases (DSE - analysis). SE and DSE calculations, called SE-DSE - analysis, provide the possibility to identify molecular descriptors more sensitively by systematic differences in databases consisting of synthetic, natural and drug substances. But at the same time, SE-DSE analyses also show that many descriptors are not sensitive to the class features of compounds, despite their complexity and high information content.

Shannon information entropy is also widely used in the study of protein sequences [49]. Information entropy is estimated here in three ways: "k-tuplet" analysis, generalised Zipf analysis and the Chou-Fasman player method. "K-tuplet" analysis is a letter analysis based on conditional sequential probabilities. Generalised Zipf analysis demonstrates the statistical linguistic qualities of protein sequences and uses word frequency in determining Shannon entropy. " Chou-Fasman" player is an algorithm based on Chou-Fasman rules for protein structure. It uses both sequence and secondary structural information to determine the number of possible amino acids that can be approximately substituted in a sequence. Using these entropy values, the number of most probable protein sequences is calculated. Typically, the number of most likely protein sequences is much smaller than the number of possible sequences.

1.3.1. Application of information theory to the development of a mathematical model of molecular complexity.

Based on the concepts of graph theory and information theory, Bertz developed a mathematical model of molecular complexity [50]. All molecules consist of atoms of some number of elements and are distinguished by the elements of which they are

composed and by the way atoms are bound to each other, i.e., by their topology. Stacked in two-dimensional space, a graph is a rather poor canonical model of a molecule. This representation is poorer compared to spatial models of molecules. However, the exact bond lengths and angle magnitudes are not important when considering complexity, what is important is the equivalence or non-equivalence of atoms, i.e. symmetry, which is most conveniently conveyed in the framework of elementary information theory.
The application of graph theory turns into the problem of finding an invariant (a quantitative value that will not change under different graph stacking) that increases whenever the number of topological features increases. Gordon and Kennedy [50] proposed an index, namely the number of ways in which propane can be "cut" from a molecule, i.e., the number of connected subgraphs of the molecular graph that are isomorphic with the propane graph. This index has only been applied to consider branching in saturated systems.
Bertz found that this index can also be directly applicable to cyclic structures [50].
To extend the above index to multiples, Bertz gave another definition of the index - the number of pairs of adjacent bonds, i.e., the number of lines in the line graph (or "bond graph") of a molecular graph. Then, using the bond graphs of various systems with multiple bonds, the number of pairs of adjacent bonds can be established.
First of all, for brevity, it is defined that a pair of adjacent links is a link ip The value of ts for a multiple link is equal to the number of lines in the multiple link simultaneously counted twice: *ts (1 2)d(d-1).*
The same formula is valid for the number of bonds of a saturated atom with degree (valence) *d*. Another definition of // is the number of tropes of length 2, i.e. simple chains of length 2. All lines in a trail must be distinct, and the vertices need not be distinct.
The symmetry of a molecular graph can now be introduced by substituting into Shannon's formula the sets of equivalent mappings *TS,* per each set *i*:

$$H = -\sum_i p_i \log_2 p_i = -\sum_i \frac{\eta_i}{\eta} \log_2 \frac{\eta_i}{\eta} \qquad (1.2)$$

A more convenient form is obtained by multiplying equation (1.2) by η, , which gives information on equivalent bundles on a "per molecule" rather than "per bundle" basis:

$$I = \eta H = \eta \log_2 \eta - \sum \eta_i \log_2 \eta_i \qquad (1.3)$$

Although H and I reflect the symmetry available for a given number of bonds (e.g., for a set of isomers), information about the size of the molecules is lost when going from η to I(η). Thus, for example, I = 0 whenever all bonds are equivalent, which is the case for such fundamentally different molecules as n-butane, cyclobutane, cyclopentane, tetrahedran, dodecahedran, and many others.
To make equation (1.3) applicable in the general case, information about the size of the molecules (number of bonds) must be added to it. This can best be done by adding $\eta \log_2 \eta$ to expression (1.3), which gives equation (1.4), reflecting both the size (through the number of bonds) and the symmetry (based on equivalent bonds) of the molecular graph.

$$C(\eta) = 2\eta \log_2 \eta - \sum \eta_i \log_2 \eta_i \qquad (1.4)$$

Let us consider the application of equation (1.4) to the example of tert-butylacetylene, which has four sets, each containing three equivalent bonds. Its complexity, therefore, is equal to: $C(\eta)=2\cdot 12\cdot \log_2 12 - 4\cdot 3 \log_2 3 = 71{,}8.$

Heteroatoms can also be included in such a scheme using the Shannon formula. Labelling a molecular graph in such a way that previously equivalent vertices become non-equivalent increases the value of C(η), since previously equivalent bundles by necessity become non-equivalent. On the other hand, the value of C(η) does not change when the labelling does not affect the equivalence classes of the mappings. Nevertheless, by being labelled, the molecule becomes more complex, and this complexity can be calculated using equation (1.3). By adding this value to equation (1.4), the value of total complexity can be obtained [50].

This mathematical model of molecular complexity has considerable generality and can be applied to some synthesis problems, in particular for the classification of synthesis reactions [50].

According to the formalism developed by Sarkar, Beysak, and Rauhaudhuri, a whole molecular graph was used to determine various graph theoretic-information invariants; the equivalence relation was defined on the set of vertices *V(G)* in such a way that two vertices belong to a given equivalence class if they have the same multiplicity of edges and the same number of first-order neighbours with the same degrees [22]. Neither the chemical identity of vertices nor the type of binding of higher-order neighbours (elements located at distance *2, 3,..., p* from the selected vertex) were considered.

1.3.2. Calculating indices based on the symmetry of vertex neighbourhoods in a chemical graph.

Magnuson et al[51] proposed to calculate topological indices based on the symmetry of vertex neighbourhoods in a chemical graph. The approach used in obtaining these topological indices is to partition the vertices of a complete molecular graph into non-overlapping subsets based on an equivalence relation defined with respect to different degrees of neighbourhood symmetry, construct a probabilistic scheme and finally calculate the amount of information using Shannon's formula.

The information content of a graph with respect to a neighbourhood of k-th order (IC_k), which is the information content per vertex, is calculated using equation (1.1). The total information content (TIC_k), i.e., the complexity measure per graph, and other derived indices, structural information content (SIC_k), binding information content (BIC_k) and complementary information content (CIC_k) are calculated as follows:

$$TIC_k = n \times IC_k,$$

$$SIC_k = IC_k / \log_2 n,$$

$$BIC_k = IC_k / \log N,$$

$$CIC_k = \log_2 n - IC_k,$$

where *n is the* power of the set V(G),
N is the total number of edges (covalent bonds) in the molecular graph.
Movšović [22] noted that the information content of graphs can be viewed as quantitative measures of their relative complexity. For a neighbourhood of a given order IC_k is maximal when all vertices are distinct. On the other hand, for the most symmetric graph (in which all vertices are equivalent), IC_k is zero. CIC_k is equal to log_2 n if the graph is most symmetric with respect to the k -th order neighbourhood, and goes to zero if all vertices are distinct. Since the above indices are obtained by considering neighbourhoods of vertices of chemical graphs, they are called neighbourhood symmetry indices.
The usefulness of such indices was demonstrated in the study of quantitative structure-activity correlations of alcohol solubility, inhibition of microsomal para-hydroxylation of aniline by cytochrome P_{450} and barbiturate toxicity [51].
Rudkevich et al. [52] made an attempt to find numerical indices of condensed coal tar aromatic hydrocarbon molecules that can be used to encode structural information when forming databases for computers. Since the most important structural characteristic of aromatic hydrocarbons, which explains their physicochemical features quite completely, can be the total number of resonance (Kekulian) structures, in the form of which the formulas of hydrocarbons are written down, the Shannon expression for the information entropy of a molecule of any aromatic compound can be represented in the form:

$$H = \log_2 N,$$

where *N is the* number of resonance structures describing the mobility of the electron cloud in the hydrocarbon molecule.
Satisfactory linear correlations were found between *H* and experimentally obtained resonance energies, as well as melting and boiling temperatures of individual polycyclic arenes. I.e., the proposed numerical index of coal tar hydrocarbon molecules (information capacity) may well be used to predict the indicated physical and chemical constants of still unstudied compounds.
The information approach using the Shannon entropy of information is probabilistic. However, non-probabilistic approaches to the definition of information are also developing, for example, Kolmogorov's approach based on the idea of epsilon-entropy (combinatorial approach) and algorithmic determination of the amount of information, the approach of Ingarden and Urbanek, etc. [53]. [53].
1.3.3. Calculation of geometric information index of order and disorder.
S.A. Dembovsky [53] and P.A. Kozmin [54, 55] proposed to obtain a new geometric information molecular index by the non-probabilistic method. A model for order and disorder in the arrangement of atoms in molecules based on Eq:

$$Q = 1 - P/_{3n},$$

where Q is the index of order in the arrangement of atoms in the molecule,
P is the number of independent coordinates (positional parameters) fixing a group consisting of *n* points in an arbitrary coordinate system, the value $P/_{3n}$ is the index of disorder.
The value of Q depends on various structural parameters of the molecule: number of

atoms in the molecule, symmetry, dimensionality, number of degrees of freedom, i.e. Q [and $(1- Q)$] is a function of several parameters $Q = f(M,n,D,CCC)$, where M is the symmetry of the point group of the molecule, measured by the multiplicity of the common position, n is the number of atoms, reflecting a measure of the complexity of the molecule (simple two- and three-atom or complex molecules), D *is* the dimensionality of the molecule, serving as a measure of physical heterogeneity, since the dimensionality is related to the anisotropy of the molecule, CCC is the number of structural degrees of freedom.

The order index Q is very sensitive to static and dynamic isomerism and correlates well with symmetry: Q decreases with decreasing symmetry, but the order remains unchanged if the multiplicity of the common position does not change. The appearance of new types of motions in non-rigid molecules (libration, permutations, and inversions) is accompanied by a decrease in symmetry and a decrease in the order index.

Let us analyse the parameters of the function from the point of view of information theory. As already mentioned, in accordance with Ashby's definition [26], information is a measure of system diversity, and in accordance with Glushkov's definition [27] - a measure of non-homogeneity and heterogeneity of the system. Symmetry in the framework of the considered approach can be considered as a measure of structural diversity or structural heterogeneity, since symmetry operations (symmetry axes and planes and inversion centre) make it possible to divide the set of atoms composing a molecule into automorphic groups (AMGs) [53]. Let us explain this on the example of the molecule CH F_{22} of symmetry C_{2v} . It has three AMGs: two of them [H(1) and H(2)] and [F(1) and F(2)] are united by a second-order axis, and the third (the C atom) lies on this axis. In this case, there is an AMG distribution of {2,2,1} symmetry. The atoms in the groups are identical (homoatomic), and the groups themselves are collectively heteroatomic. Thus, the symmetry reflects the structural heterogeneity and indirectly the chemical diversity of the molecule. I.e., the value Q characterises the molecule quite fully and qualitatively corresponds to the definition of information as a measure of diversity, heterogeneity and non-homogeneity.

Correlations were found between the parameter $P/3n$, characterising disorder and the information symmetry index I_{sym} according to Bonchev [53 - 55]. The calculation of I_{sym} takes into account the distribution of symmetry-equivalent atoms, which include atoms that change places under the action of symmetry operations [56 - 57]. Each symmetry operation covers a certain number of atoms of the molecule, and also their certain groups - AMGs.

Thus, the value of $P/3n$, can be considered as a special information index - geometric molecular information index

$$I_{geom} = P/3n$$

Its similarity to the symmetry index I_{sym} is due to the fact that both indices take into account the symmetry of a molecule: I_{sym} by means of the corresponding AMGs, and

$P/_{3n}$, - by geometrical consideration of a given molecule. The difference is explained by the difference not only in the way symmetry is taken into account, but also in the method of calculation, which is probabilistic (according to the Shannon formula) for I_{sym} and non-probabilistic (only from the considerations of analytical geometry) for the $P/_{3n}$,
Suppose that a degree of freedom associated with chain unfreezing, or isomerism, etc., has appeared in the molecule. In this case, the value of $P/_{3n}$, will grow, and hence Q will fall: the information content increases as the order index of Q decreases. Then, if I_{geom} can be compared with *Isym* and through it with information entropy according to Shannon, then the index of order Q can be compared with non-entropy according to Brillouin [17] or information redundancy.
The limiting case Q =1 corresponds to a regular polygon with an infinite number of sides, i.e. a circle or an infinite chain with fixed CCCs. Real molecules always have an information content $P/_{3n}$, and the greater the closer to the beginning (Q =0) of the order scale they are located.
The I index$_{geom}$ can be used to establish composition-structure-property relationships, e.g., to estimate entropy, enthalpy, reactivity, to predict the properties of molecules, to guide synthesis of molecules with given properties, to systematise for computerisation of research, etc.
Thus, it is evident from the above that the calculation of various information indices quantitatively characterising the structure of molecules plays an important role in the encoding of structural information and allows the optimal formation of appropriate databases. This, in turn, opens wide prospects for prediction of substances with given properties.
In this paper we have tried to investigate a rather fundamental problem: How the structure of space changes when a molecular system is placed in it. In other words, we tried to model the information field of a molecule. But, it should be noted that unlike the anthropomorphic approach of academician Berg, according to which only a human being creates an information field around himself [14], and various parascientific "theories" of the information field, we postulate that any object creates an information field around itself, and the following sections of this book are devoted to the consideration of the main provisions of the developed concept of the information field.

CHAPTER 2

MOLECULAR INFORMATION FIELD

This section outlines the main provisions and principles of the concept of the information field (IF) of molecules developed by us [58-67]. The FE concept is based on the consistent application of the Shannon approach not only to the object - molecule, but also to the surrounding space. That is, the space surrounding the object is discretised and considered as a system of elements. At each point of such space, using the Shannon approach, one can calculate the FE parameters of the molecule.

This model describes the situation when an object (molecule) structures the space around it, i.e. generates information in it.

2.1. Fundamentals of the concept

In Section 1.1. it was noted that in 1948 K. Shannon proposed a model that allows quantifying the information contained in a message. K. Shannon defined information as information that removes the uncertainty existing before its receipt [16]. The degree of uncertainty of a situation can be characterised by entropy. Thus, the degree of uncertainty of a situation can be related to the number of possible variants (paths) of development of this situation (outcomes of experience). Hence, it is clear that the objective basis of information is "difference" or "diversity", which is consistent with the information concepts of W.R. Ashby and V.M. Glushkov [68]. In accordance with Hartley's hypothesis [7], the logarithm of the possible number of outcomes (paths) of the analysed situation is used to estimate the amount of information. If we use the logarithm on the base 2, then the removal of uncertainty of the situation is equal to 1 ($I = log_2 2 = 1$). This corresponds to the amount of information in one "bit". If different outcomes (i) of the situation development have different probabilities (Pi), the amount of information is more correctly estimated using the Shannon formula.

Shannon's approach is based on a discrete representation of an "object" (situation, system) consisting of a set M of "elements" (outcomes). Let the power of this set be equal to n ($|M| = n$). If we define an equivalence relation in this set, then all its elements form a number of subsets (mj), each of which includes only equivalent elements

$$m_1 \cup m_2 \cup \cdots m_i \cdots \cup m_k = M$$

$$k \leq n \qquad m_j \cap m_i = 0 \qquad |m_i| = n_i \qquad \sum_i n_i = n$$

The probability of choosing any element (situation outcome) from the *i-th* subset is equal to $pi = ni/n$. According to Shannon, the amount of information per one element of the system (object) is defined as

$$I = -\sum_i^k p_i lb(p_i) \qquad \text{where } lb = log2$$

For the previously mentioned situation with two equally likely outcomes (e.g., tossing a coin "heads", "tails"), just as by Hartley, we will get 1 bit of information.

$$I = -1/2\, lb(1/2) - 1/2\, lb(1/2) = 1$$

The information contained in the whole set (*M*) of elements can be easily defined as

$$I' = n \cdot I = -\sum_i n_i lb(n_i/n)$$

Hence, the piece of information defined by one of the subsets *mi of* the set *M* is equal to -*nilb(ni/n)*, and the information generated by one element of such a subset is equal to -*lb(ni/n)* or *lb(n/ni)*, respectively.
If there is no uncertainty, i.e. the situation has only one outcome and the system consists of elements of only one kind (no diversity), *then I'=I=0*, since $M = m1$; $n = n_1$ and $p1 = 1$.
We have maximum information for the case when the situation has the maximum possible number of outcomes and the system consists of only different elements: $ni = 1$; $pi = 1/n$; $I = lb(n)$; $I' = nlb(n)$.
Thus, the information content of the system is determined only by the number of elements (n) in it and the scheme of their partitioning into subsets (*jm.il* = n_i). This approach completely ignores the qualitative side of information, its semantic content. The positive side of this is the possibility to analyse any information processes and systems from general positions. In fact, a universal language for describing the information side of any process, phenomenon or object is proposed.
In this paper we have tried to investigate a rather fundamental problem: "What is the result of transferring information about the molecular structure to the surrounding space?". In other words, we tried to model the information field of a molecule. It turned out to be not so difficult to construct a formal FE model. For this purpose, it is necessary to consistently apply the Shannon approach not only to the investigated "object", but also to the surrounding space. In other words, this space, as well as the "object", should be discretised and considered as a system of elements. It should be noted that it is more correct to speak not about IP, but about the potential information field (PIF), since information arises only when there is a "receiver" ("observer") perceiving it. For the sake of simplicity, we will further use the term IP, at the same time implying PPI.
Let us imagine an idealised situation - an "empty" infinite space, (two-dimensional for simplicity), divided into identical areas (cells). It contains 0 bits of information, because all cells are exactly the same, there is no "heterogeneity" (diversity), hence no information. But as soon as we put any object into this space, the equivalence of many cells disappears. They become different, depending on how differently the object is "seen" from this or that cell of space. That is, being in different places of space surrounding the object, we (receiver, observer) can get different amount of information about this object. Thus, in order to model the IP of an object in the selected area we must learn to differentiate the cells of the space surrounding it. When all cells in the investigated area of space will be divided into groups, we will only have to apply Shannon's formula and calculate the potential information contained in each cell, group of cells and in the whole area.
Let us consider the simplest example. Let us place an object representing a single cell (point) on the plane, then the two-dimensional space around it will be divided into groups of equivalent cells forming concentric layers surrounding the object (Fig. 2.1(a)). Next, we apply the Shannon approach sequentially. Cell *A is* unique and contains maximum information. As we move away from cell *A,* the potential information decreases.

According to the formula, the information potential in a given cell depends only on the total number of cells under consideration (n) and the number of cells of a given group (ni)* . Hence it is clear that in each subsequent layer the information potential is smaller than in the previous one. In the limit, the cells of space form concentric circles for the 2-D model (Fig. 2.1(6)), or spheres for the 3-D model.
Thus, the considered FE model correctly reflects the natural decrease of potential information as the "observer" moves away from the object.
To apply this FE model to more complex objects consisting of several elements, it is necessary to take into account that information, as well as entropy, has the property of additivity. This means that we can model the FE of a complex object as a superposition of the FE of its elements. Moreover, using any properties of the object elements as weight characteristics, we can construct weighted FEs. In fact, each such field reflects information about the distribution of the considered property in space. For example, Fig. 2.2 shows the IP weighted by atom masses for the benzene molecule.
Summarising the obtained results, we can conclude that the model of the potential information field from a formal position describes the situation when an object structures the surrounding space, i.e. generates potential information in it.

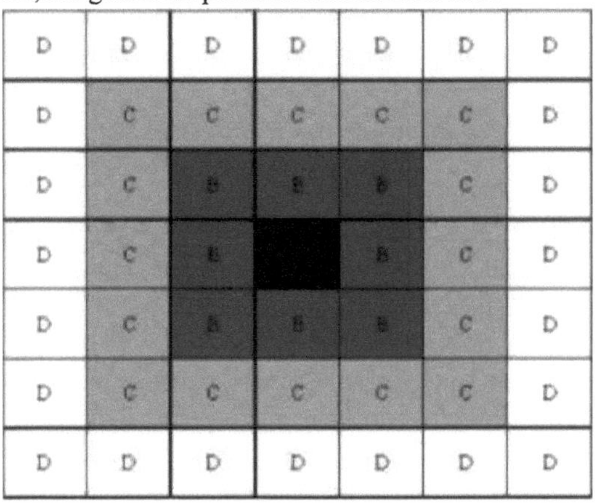

$1(A) = lb(N/T)$, $1(B) = lb(N/8)$, $1(C) = lb(N/16)$, $1(D) = lb(N/24)$
N is the number of cells of the investigated area of space.
a)

* In the illustrated model under consideration, we do not distinguish "corner" cells from "non-corner" cells for simplicity."

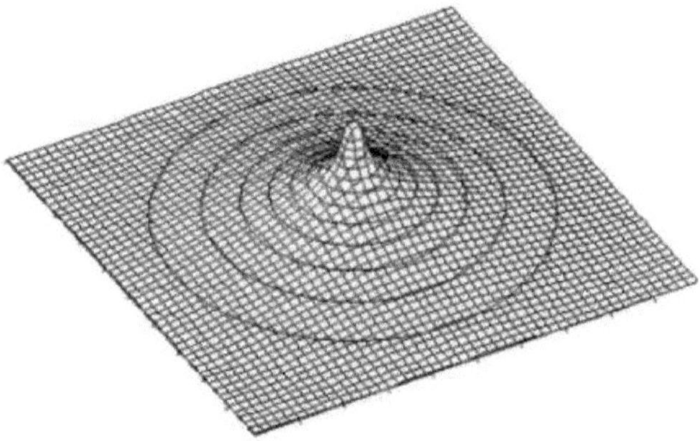

б)

Fig. 2.1. Fragment of information field of point object for 2D model.

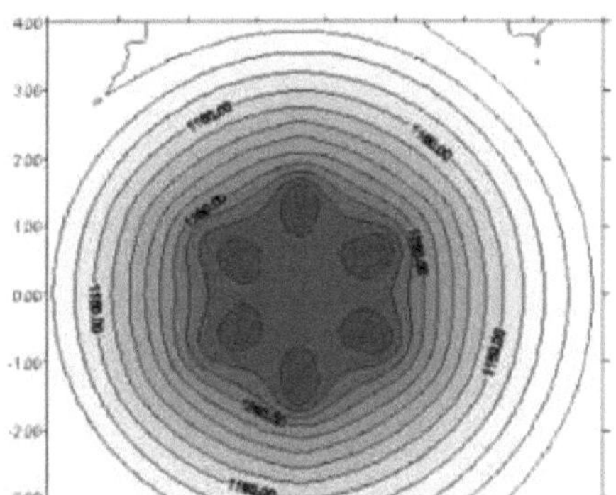

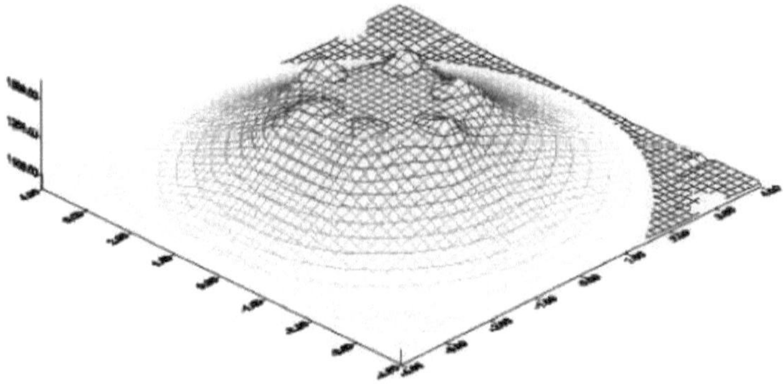

Figure 2.2: IP of benzene weighted by atomic masses.

Due to what interactions such structurisation occurs? In all probability, due to the complex of real physical fields of the investigated object. However, in this context we do not consider how space structurisation occurs, but only analyse its result. We do not reject physical mechanisms of information transfer from the object to the surrounding space, we just do not specify them.

2.2. Modelling principles

Developing this approach, it is easy to pass from the lattice model to the isotropic model of the information field for spaces of any dimension. Let us illustrate the above for a two-dimensional space. Let the discretisation step of the space be equal to ε , where $\varepsilon \to 0$ is the tuning parameter of the model. To estimate the potential of the information field on the plane (2 D - approach) for a point object at a distance r from the source (*I(r)*), we need to calculate the binary logarithm of the ratio of the area ratio of the whole area of the investigated area (*I(r)*).

space (*S^*) to the area of a circular ring (*S,.,,*) of thickness *s*. I.e.:

$$I(r) = lb(\frac{S_{общ}}{S_{сл}}) = lb\left[\frac{\pi R^2}{\pi((r+\varepsilon)^2 - r^2)}\right] = 2lb(R) - lb(\varepsilon) - lb(2\cdot r + \varepsilon).$$

Similarly, we calculate *I(r)* for three-dimensional space, only here we consider the binary logarithm of the ratio of the volume of the whole region of the investigated space (*Uo bsh.*) to the volume of a spherical shell ($U_{сл}$) with thickness *e*. I.e.:

$$I(r) = lb(\frac{V_{общ}}{V_{сл}}) = lb\left[\frac{\frac{4}{3}\pi R^3}{\frac{4}{3}\pi((r+\varepsilon)^3 - r^3)}\right] = 3lb(R) - lb(\varepsilon) - lb(3r^2 + 3r\varepsilon + \varepsilon^2).$$

The whole procedure for spaces of different dimensionality is illustrated in Fig. 2.3 (here: *R* - defines the whole space region under study (*r* varies from 0 to *R*); *I(r)* - information potential of the space region at a distance r from the "source" of the field; *I(0)* - information potential at the point corresponding to the field source; *I(R)* - information potential at the boundary of the space region under study).

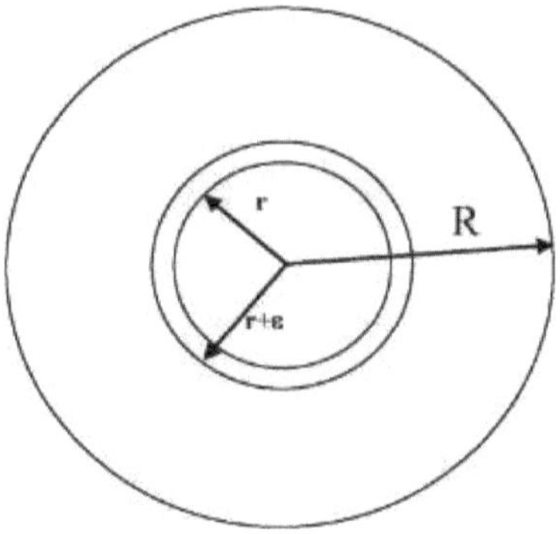

2D - space:

$I(r) = lb[(\pi R^2)/(\pi((r+\varepsilon)^2 - r^2))] = 2lb(R) - lb(\varepsilon) - lb(2r + \varepsilon)$;
$I(0) = 2lb(R) - 2lb(\varepsilon) = 2lb(R/\varepsilon)$;
$I(R) = 2lb(R) - lb(\varepsilon) - lb(2R+\varepsilon) \approx lb(R) - lb(\varepsilon) - lb(2)$;
$\Delta I = I(0) - I(R) \approx lb(R) - lb(\varepsilon) + lb(2)$

3D - space:

$I(r) = lb[((4/3)\pi R^3)/((4/3)\pi((r+\varepsilon)^3 - r^3))] = 3lb(R) - lb(\varepsilon) - lb(3r^2+3r\varepsilon+\varepsilon^2)$;
$I(0) = 3lb(R) - 3lb(\varepsilon) = 3lb(R/\varepsilon)$;
$I(R) = 3lb(R) - lb(\varepsilon) - lb(3R^2+3R\varepsilon+\varepsilon^2) \approx lb(R) - lb(\varepsilon) - lb(3)$;
$\Delta I = I(0) - I(R) \approx 2lb(R) - 2lb(\varepsilon) + lb(3)$

ND - space:

$I(0) = (N)lb(R/\varepsilon)$;
$I(R) \approx lb(R) - lb(\varepsilon) - lb(N)$;
$\Delta I \approx (N-1)lb(R) - (N-1)lb(\varepsilon) - lb(N)$

Fig.2.3. Isotropic model of the information field.

When the dimensionality of space increases, there is a rapid dissipation of the potential (Fig. 2.4). I.e., the higher the dimensionality, the faster the reduction of the FE potential occurs.

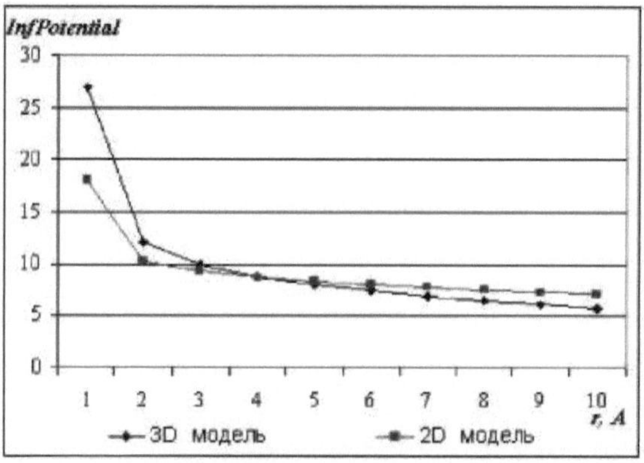

Fig.2.4. Variation of information potential of a point object with distance from the

information source for 2D and 3D models.

For the 1D model, the information potential is a constant. In this case there is no damping effect because there is no information dissipation.
It should be noted that all subsequent calculations of information parameters were carried out within the framework of the isotropic FE model for three-dimensional space.

2.3. Calculation of parameters of the information field of molecules

The term "field" is usually used in physics to denote a part of space (or the whole space) in which some physical phenomenon is considered. For example, air temperature at different points of space forms a field of temperatures, and atmospheric pressure forms a field of pressures. An electric charge creates an electrostatic field around itself: a force, quite definite in magnitude and direction, acts on each electric charge placed in some point of the field.
Many examples of this kind can be given. In all cases when we are talking about a phenomenon characterised by a scalar quantity (temperature, pressure, etc.), the field is called scalar. If the phenomenon under consideration is characterised by a vector quantity, the field is called vector.
Similarly, we can give a mathematical definition of a field [68, 70]. If each point of space M corresponds to a scalar quantity U, then there exists a scalar field $U(M)$ (e.g., the field of temperature, density in an inhomogeneous medium). If a point M has Cartesian coordinates (x, y, z), then we also write $U=U(x, y, z)$ or $U=U(r)$ *with a* vector argument (radius-vector $r=xi+yj+zk$). The properties of scalar fields can be visualised using level surfaces. These are surfaces in space on which U has a constant value. They are described by the function $U(x, y, z)=const$. Since the function defining a scalar field is often called a potential, regardless of its physical meaning, level surfaces are also called equipotential surfaces, i.e. surfaces of equal potential. *It* may be added that the value U characterising the scalar field may also depend on time, for example, the air temperature at the same points may be different at different moments of time. However, we will limit ourselves to consider only such a scalar field where U does not depend on time. Such a field is called stationary.
If each point M of space corresponds to a vector V, we speak of a vector field $V(M)$ (e.g., the velocity field of particles of a moving fluid, the field of electric tension, the field of magnetic tension, etc.). As in the case of scalar fields, in Cartesian coordinates it is written: $V = V(x,y,z)$ or $V=V(r)$ (r- radius-vector). A vector field can be written in Cartesian coordinates in the following form: $V(x, y, z)= V_x (x, y, z)i+ V_y (x, y, z)j+ V_z (x, y, z)k$. The components V_x , V_y , V_z form three scalar fields and uniquely define $V(r)$, *the* vector function of the vector argument. As above, here also we will restrict ourselves to consider only a stationary vector field independent of time.
It should be noted that in our case the information field is not a physical field but a mathematical field.

2.3.1. A model of a scalar information field.

In the framework of this model, each point of space corresponds to a scalar quantity - the FE potential. In Section 2.2. the formulas for calculating the FE potential for spaces of any dimensionality are given. By introducing weight parameters reflecting any property of

the object elements, it is possible to calculate FE potentials weighted by this property. Since UI is a scalar field, we can calculate for it the gradient[2] *(bit-A'l*) and divergence (*bit A'3*). The field gradient is a vector which, in direction and numerical value, characterises the largest rate of change of the scalar quantity *U* and is defined at each point of the field _ _ . . . *∂i., di ., di.*
by a ratio:

$$gradU = \frac{\partial U}{\partial x} i + \frac{\partial U}{\partial y} j + \frac{\partial U}{\partial z} k .$$

Thus, the IP gradient is a vector quantity that reflects the direction of maximum change in the information field. It should be noted that we calculate the value of the IP antigradient (gradient with a negative sign), since it is important for us to analyse how information spreads from the source to the surrounding space, and not vice versa.
The divergence $divV^2$ expresses the specific density of sources or drains of the field[3] . If $divV = 0$ in a region, then the vector field *V(M)* in that region is called source-free. Divergence is defined as:

$$divV = \frac{\partial V_x}{\partial x} + \frac{\partial V_y}{\partial y} + \frac{\partial V_z}{\partial z} .$$

To construct the formulas for calculating the gradient and divergence of the IP gradient field, it is necessary to find the first and second derivatives of the potential, respectively. As a result, the following basic formulas for calculating the gradient and divergence of the IP gradient field were obtained:

$$GrdX = 3 \cdot (\frac{(x-a) \cdot (2d+\varepsilon)}{d \cdot (3d^2) + 3 \cdot \varepsilon \cdot d + \varepsilon^2 \ln 2});$$

$$GrdY = 3 \cdot (\frac{(y-b) \cdot (2d+\varepsilon)}{d \cdot (3d^2) + 3 \cdot \varepsilon \cdot d + \varepsilon^2 \ln 2});$$

$$GrdZ = 3 \cdot (\frac{(z-c) \cdot (2d+\varepsilon)}{d \cdot (3d^2) + 3 \cdot \varepsilon \cdot d + \varepsilon^2 \ln 2});$$

$$Grad = \sqrt{(GrdX)^2 + (GrdY)^2 + (GrdZ)^2}$$

$$Div = -3 \cdot \frac{6 \cdot d \cdot (d+\varepsilon)^2 + \varepsilon^2 \cdot 3d + 2 \cdot \varepsilon}{d \cdot (3d \cdot (d+\varepsilon) + \varepsilon^2)^2 \cdot \ln 2};$$

$$d = \sqrt{(a-x)^2 + (b-y)^2 + (c-z)^2} ,$$

where *GrdX, GrdX, GrdX* are the projections of the vector - the IP antigradient on the x, y, z axes;
Grad is the length of the vector - the IP antigradient;
Div is the divergence of the IP gradient field;

---

[2] The word "gradient" comes from Latin *gradiens* - walking. The concept of "gradient" was first used in the 19th century in meteorology as an indicator of the rate of change of atmospheric temperature and pressure.
[3] The symbol *div* comes from the Latin word *divergere* - to detect divergence.
for the considered model of the information field "drains" are absent

a, b, c - coordinates of the source;

x, y, z - current coordinates of a point in space.

In all cases, we used the model tuning parameter e = 0.05.

When solving various structure-property problems for the studied molecules, the characteristics of the potential information field in different regions of the near-molecular space can be used as structural parameters: potential, gradient, and divergence of the gradient field. In addition, these parameters, as indicated earlier can be weighted by different properties of atoms. It should be noted that the weights of atoms are taken in relative fractions, i.e., the value of a weight parameter of a certain type for a given atom is determined by the ratio of this weight parameter of the atom to the sum of the corresponding weight parameters of all atoms of the molecule. I.e., the dimensionless weight parameters of property P are determined by the formula:,

$$V_i = \frac{P_i}{\sum_{i=1}^{n} |P_i|}$$

Where i is the atom number

n is the total number of atoms in the molecule;

Pi is the atomic characteristic under consideration.

Thus, the weight parameters reflect the degree of informational influence of this or that atom.

Fig.2.5 shows for illustration the maps of potential, gradient and divergence of the IP gradient field for the molecule 1-bromo-3-fluorobenzene. In calculating these parameters, atomic mass was used as a weight characteristic.

Consider a scalar field given by the function $U = U(x,y,z)$, which is assumed to be a single-valued and continuous function of x, y and z having continuous partial derivatives of the first order. Let us define those points where the value of the parameter is the same, i.e. the points where the function $U(x,y,z)$ takes the same value C: $U(x,y,z) = C$. This equation can be regarded as the equation of some surface in space.

It seems important to define a "first" equipotential surface covering the entire molecule, since such a surface reflects information about the shape of the molecular structure. In this paper we use instead of the term "equipotential" the term "equi-informational surface", implying that we mean a surface with one value of the IP parameter (potential, gradient, or divergence of the gradient field). As an illustrative example, Fig. 2.6 and Fig. 2.7 show the "first" equipotential-information surfaces of the FE parameters, fully covering the molecular structure, for the molecules 1-bromo-2-chloro-4-fluorobenzene and dibromodifluoroethylene, respectively.

It should be noted that it is the "first" shell, which completely covers the structure, that carries important information. If we take the value of the IP parameter very large, we will get a region enclosed inside the structure, i.e. such an equi-informational shell will not completely cover the whole molecule and the information about the shape is lost (Fig. 2.8).

Potential

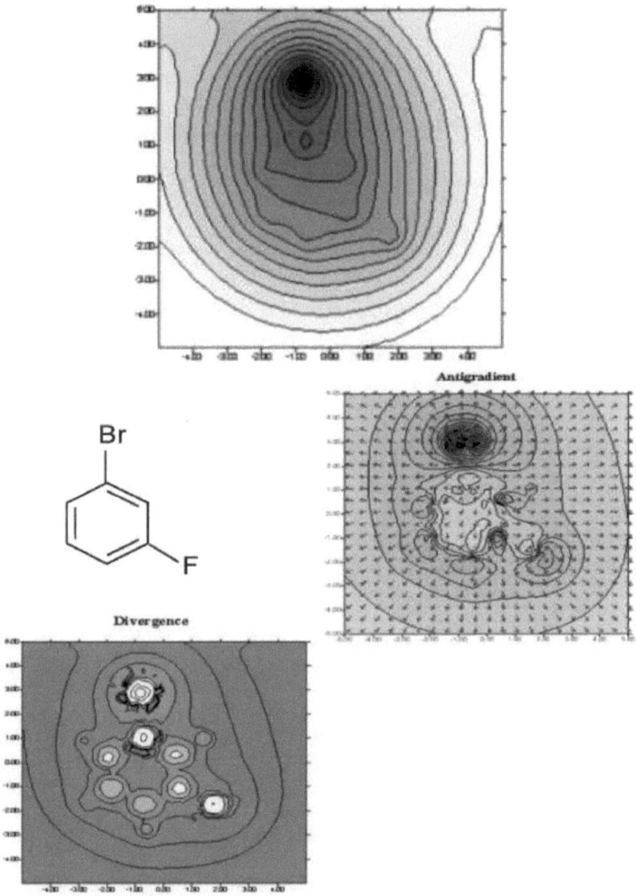

Figure 2.5. Information field parameters of 1-bromo-3-fluorobenzene.

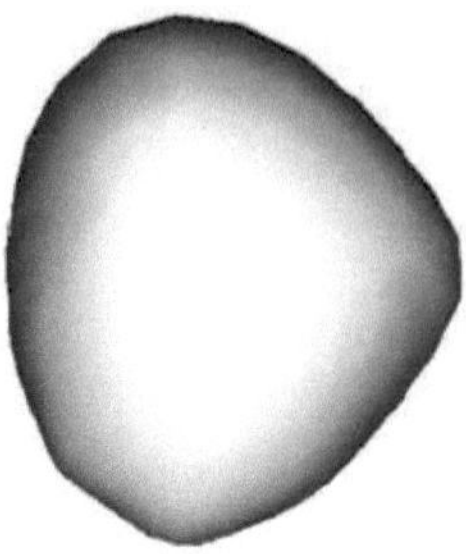

a)

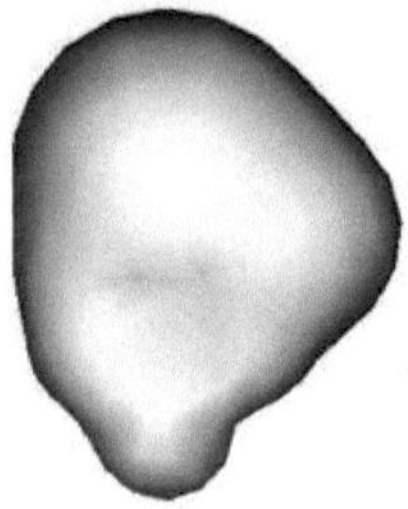

б)

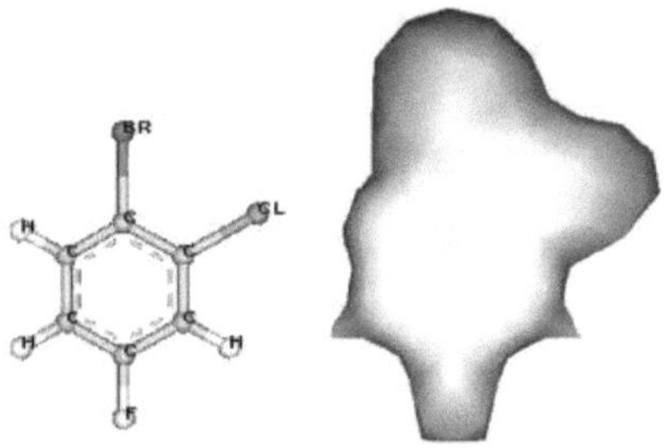

в)

Fig. 2.6: Equi-information surfaces of FE parameters for the molecule 1-bromo-2-chloro-4-fluorobenzene: a) potential, b) gradient, c) divergence.

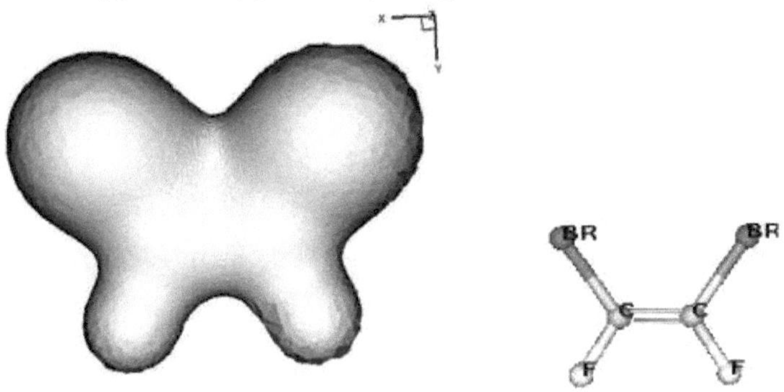

Figure 2.7. Equi-information shell of the IP divergence parameter for the Z-dibromodifluoroethylene molecule.

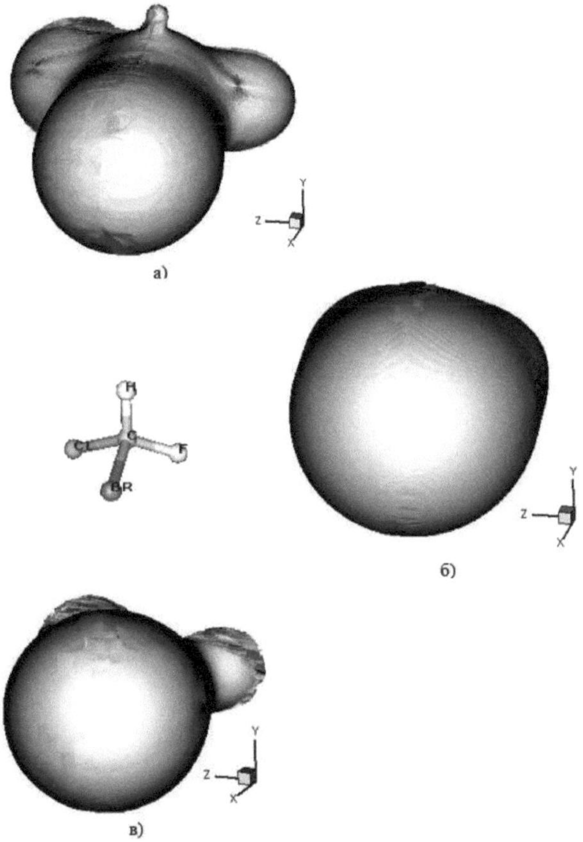

Fig.2.8. Equi-information shells of the IP divergence parameter for the bromochlorofluoromethane molecule: a) "first", b) "larger", c) "smaller".

If we take a very small value of the IP parameter, we obtain a region that encloses the investigated structure, but the details of the shape become blurred, i.e., in this case the information about the shape of the molecular structure is also lost. When constructing equi-information surfaces, it was found that the divergence parameter of the IP gradient field reflects the shape of the molecular structure in more detail compared to the potential and gradient, i.e., it contains more accurate information about the shape of the molecule. It should also be noted that weighted information fields can be constructed. As a result of the analysis of equipotential shells of various molecules, it was suggested that equipotential shells can be used in the study of the structure of a molecule, its shape.

2.3.2. Vector model of the information field.

The vector information field model is proposed as a development of the FE concept. If, when modelling a scalar FE, each point of space corresponds to one scalar quantity, when considering a vector FE, each point of space corresponds to three scalar characteristics, which are actually vector components. I.e., if an object (molecule) is represented as a

system of vectors - sources, then using the components of these vectors as weight parameters, it is possible to construct a vector FE of the studied object (molecule) (Fig. 2.9).
Let a certain vector *a* corresponds to an element of the object:

$$a(t) = a_x(t)i + a_y(t)j + a_z(t)k\,,$$

where $a_x(t),\ a_y(t),\ a_z(t)$ are the projections of vector *a* on the axes of *the* rectangular coordinate system in space, *i, j, k are the* system of orthogonal coordinates, vector *a* is a function of some property *t.*
Then at each point of the space surrounding the object, the corresponding vector of the vector FE can be calculated (Fig. 2.9).
The length of the FE vector *LV(i)* at point *i* of the space is determined by the formula:

$$LV(i) = \sqrt{(a'_x(i))^2 + (a'_y(i))^2 + (a'_z(i))^2}$$

where $a_x(i),\ a_y(i),\ a_z(i)$ · are projections of the FE vector at the current point of space on the *x-axis,*
y, z, calculated as:

$$a'_x(i) = I(i) \cdot a_x(i);$$
$$a'_y(i) = I(i) \cdot a_y(i);$$
$$a'_z(i) = I(i) \cdot a_z(i);$$

where $a_x(i)$, $a_y(i)$, $a_z(i)$ *are* projections of the IP source vector on *x, y, z* axes; *I(i) is the* information potential at point *i.*

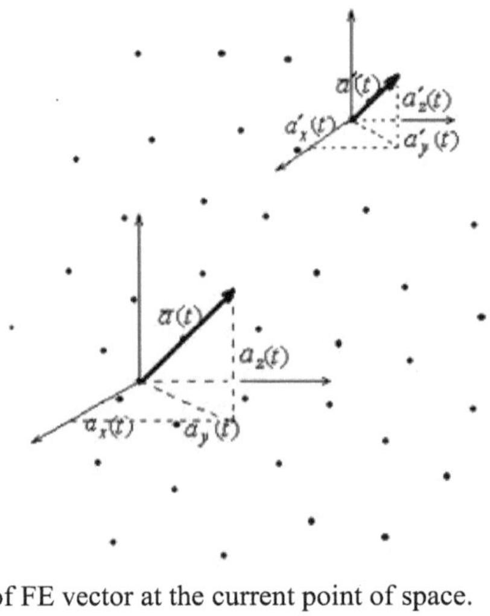

Fig. 2.9. Construction of FE vector at the current point of space.

The formulas for calculating the vector potential (*VIP(i))* and divergence (*VD(i))* of vector IP are given below.

$$VIP(i) = 3 \cdot x(i) \cdot (\frac{(x-m) \cdot (2d+\varepsilon)}{d \cdot (3d^2) + 3 \cdot \varepsilon \cdot d + \varepsilon^2 \ln 2}) \cdot a_x(i) +$$
$$+ 3 \cdot y(i) \cdot (\frac{(y-n) \cdot (2d+\varepsilon)}{d \cdot (3d^2) + 3 \cdot \varepsilon \cdot d + \varepsilon^2 \ln 2}) \cdot a_y(i) +$$
$$+ 3 \cdot z(i) \cdot (\frac{(z-p) \cdot (2d+\varepsilon)}{d \cdot (3d^2) + 3 \cdot \varepsilon \cdot d + \varepsilon^2 \ln 2}) \cdot a_{z;}(i)$$

$$VD(i) = a_x''(i) + a_y''(i) + a_z''(i),$$

Where

$$a_x''(i) = 3 \cdot (\frac{(x-m) \cdot (2d+\varepsilon)}{d \cdot (3d^2) + 3 \cdot \varepsilon \cdot d + \varepsilon^2 \ln 2}) \cdot a_x(i);$$

$$a_y''(i) = 3 \cdot (\frac{(y-n) \cdot (2d+\varepsilon)}{d \cdot (3d^2) + 3 \cdot \varepsilon \cdot d + \varepsilon^2 \ln 2}) \cdot a_y(i);$$

$$a_z''(i) = 3 \cdot (\frac{(z-p) \cdot (2d+\varepsilon)}{d \cdot (3d^2) + 3 \cdot \varepsilon \cdot d + \varepsilon^2 \ln 2}) \cdot a_z(i).$$

If the object is a system of vectors, then at each point of the
of the surrounding space, the superposition of vector IP elements is calculated. As an example, we can consider the vector information field and divergence of the vector IP of the 1-bromo-3-chlorobenzene molecule represented by two vectors whose lengths reflect the polarity of carbon-halogen bonds (Fig. 2.10.).

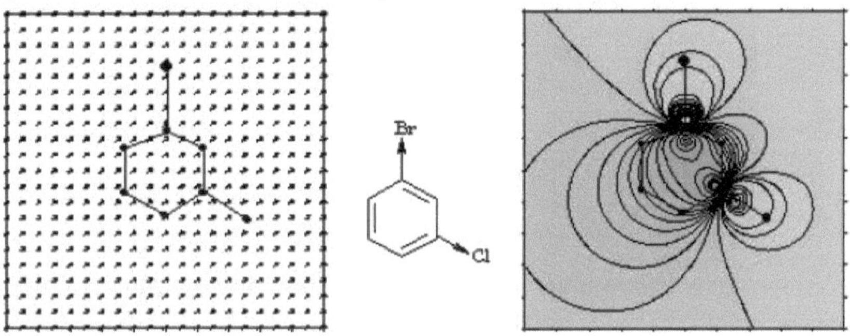

a) б)

Fig.2.10: Vector FE (a) and divergence of vector FE (b) of 1-bromo-3-chlorobenzene.

The vector model of the IP can be used when the molecule under study can be represented in the form of vectors, for example, reflecting the dipole moments of the groups composing this molecule. It seems to us that this vector information field model can also be used in the analysis of hydrogen bonds. Considering a molecule as a system of vectors of a certain length, reflecting donor and acceptor fragments of the molecule, it is possible to construct a vector IP for it, characterising the possible potential formation of a hydrogen bond (Section 4.1.1.).

Besides, it is proposed to calculate one more important parameter of the vector field - the *rotor* of the vector field *rot*[4]. If we consider the IP gradient field, it is potential and is well defined by one scalar function - potential. There is a theorem on the conditions of potentiality: for a vector field $a = a(x, y, z)$ to be potential, it is necessary and sufficient for it to be vortex-free, i.e., to be potential.

$$rot\ a = 0.$$

If we consider the model of a vector information field, then for such a field it makes sense to calculate a rotor, since such a field is no longer potential.

Within the framework of the vector FE model, a molecule is represented as a system of vectors. In addition to the above proposed ways of vector representation, we used one more scheme of representation of vectors characterising the object under study. Each atom corresponds to a certain value of potential weighted by some property of atoms. Considering each bond in a molecule, we can generate for each atom forming this bond a vector whose length is equal to the difference of potentials in the corresponding atoms divided by 2 (Fig. 2.11). The resulting vector is directed along the bond, from the atom with the larger potential value to the atom with the smaller value. As a result of this procedure, when considering all bonds in a molecule, for each atom we have a set of vectors, which are then added up and the resulting vector is obtained. To illustrate this vector calculation scheme, an example for an object consisting of three elements is shown in Fig. 2.11. As a result, for each element we obtain vectors $a1, a_2$, $a3$ respectively.

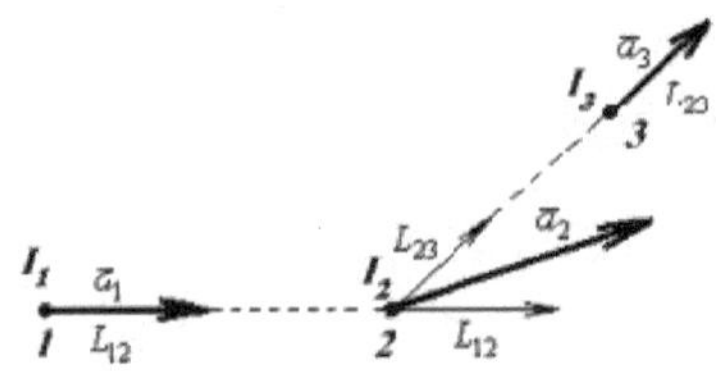

$$I_1 > I_2\ ; \qquad L_{12} = \frac{|I_1 - I_2|}{2}\ ;$$
$$I_2 > I_3\ ; \qquad L_{23} = \frac{|I_2 - I_3|}{2}\ .$$

Fig. 2.11. Example of forming a vector model of an object consisting of three elements

Having obtained by the above method the system of corresponding vectors, it is possible to calculate in each point of the surrounding space the rotor - vector with *dA*, projections $\frac{\partial A_z}{\partial y} - \frac{\partial A_y}{\partial z}, \frac{\partial A_x}{\partial z} - \frac{\partial A_z}{\partial x}, \frac{\partial A_y}{\partial x} - \frac{\partial A_x}{\partial y}$ on coordinate axes.

[4] The symbol *rot* comes from the abbreviated Latin word *rotatio* - rotation. The symbol *curl* also comes from the English word *curl - curl, curl.*

$$\begin{cases} rot_x A = \dfrac{\partial A_z}{\partial y} - \dfrac{\partial A_y}{\partial z}; \\ rot_y A = \dfrac{\partial A_z}{\partial z} - \dfrac{\partial A_z}{\partial x}; \\ rot_z A = \dfrac{\partial A_y}{\partial x} - \dfrac{\partial A_x}{\partial y}. \end{cases}$$

We get it:

$$rotA = (\frac{\partial A_z}{\partial y} - \frac{\partial A_y}{\partial z})i + (\frac{\partial A_x}{\partial z} - \frac{\partial A_z}{\partial x})j + (\frac{\partial A_y}{\partial x} - \frac{\partial A_x}{\partial y})k.$$

Or in another post:

The length of the rotor-vector is determined by the following formula:

$$Rotor = \sqrt{(rot_x A)^2 + (rot_y A)^2 + (rot_z A)^2}.$$

A parameter such as the rotor of a vector FE can also be used as a structural parameter to solve QSAR problems.

2.4. Influence of changes in the structure of a molecule on the parameters of its information field

To illustrate the possibilities of the developed information field concept, a number of model examples - various benzene derivatives - were considered.

Within the framework of the scalar FE model, such FE parameters as potential, gradient, and divergence of the FE gradient field were calculated for all investigated systems. When calculating these parameters, the mass of atoms of molecules was used as a weight characteristic. The investigated molecules were placed entirely in a region of space representing a sphere of radius 5A. Then, having at each point the values of the above-mentioned IP parameters, we calculated for each molecule the total information potential (*SumI*), the information potential in occupied cells, the average characteristics (I) of the parameters and their standard deviation (S):

$$SumI = \sum_{i=1}^{N} I_i; \quad I = \frac{\sum_{i=1}^{N} I_i}{N}; \quad S = \sqrt{\frac{\sum_i (I_i - I)^2}{N-1}}.$$

where N *is the* total number of points in the space;

I_i - value of the parameter at point i.

Monosubstituted benzenes were considered in the first sample of investigated molecules. As can be seen from Fig. 2.12, the total information potential and the average characteristics of the information fields in all cases naturally increase with increasing substituent mass. The standard deviation of the mean square deviation of the various characteristics of the information field behaves in a similar way.

In the second sample of investigated molecules, all isomers of difluorobenzene were considered (Fig. 2.13). As one would expect, all the calculated characteristics of the information field naturally decrease in the transition from the o-isomer to the p-isomer. Obviously, this is due to the mutual influence of the information fields of fluorine atoms. For the o-isomer, fluorine atoms maximally enhance the mutual influence of each other.

The most interesting sample of investigated molecules are all isomers of fluorochlorobromobenzene. If we analyse the information content of the molecular graph

for all of them, we can easily see that it is the same for all compounds and amounts to the value lb(9) (taking into account the environment, all atoms are different).
However, the characteristics of the information field for all these molecules are different and allow us to discriminate all these molecules (Figure 2.14).
The synergism of mutual influence of the information fields of substituents is maximal for isomers 1, 2, 4. For isomers 7, 8, 10, the information interaction of substituents is minimal, since they are the most distant from each other.
This sample of fluorochlorobromobenzene isomers was also studied within the framework of the vector FE model (Fig. 2.15). The molecules were represented as vector systems obtained according to the scheme outlined in Section 2.3.2. The potential, divergence and rotor of the vector FE were calculated for these compounds. When calculating the vector FE parameters, the mass of molecule atoms was used as a weighting characteristic, as in the case of scalar FE. The investigated molecules were entirely placed in a region of space representing a sphere of radius 5A. At each point of the selected region, the above parameters of the vector IP were calculated. Then statistical processing of the data was carried out, the mean characteristics of the parameters (*I*) and their standard deviation (S) were calculated for each molecule.
As a result of the analysis of the obtained data using the vector IP model, it is seen that in this case the obtained structural parameters also allow us to distinguish isomers. If the substituents are the most distant from each other, minimal information interaction is observed (isomers 7, 8, 10). The synergism of mutual influence of information fields of substituents is maximal for isomers 1, 3, 4, because heavy substituents in these molecules are located closer to each other than in other isomers.
Thus, it can be seen that the information field parameters are quite sensitive to changes in the molecular structure and, in all probability, can be used as structural parameters to solve "structure-property" problems.

Information capacity

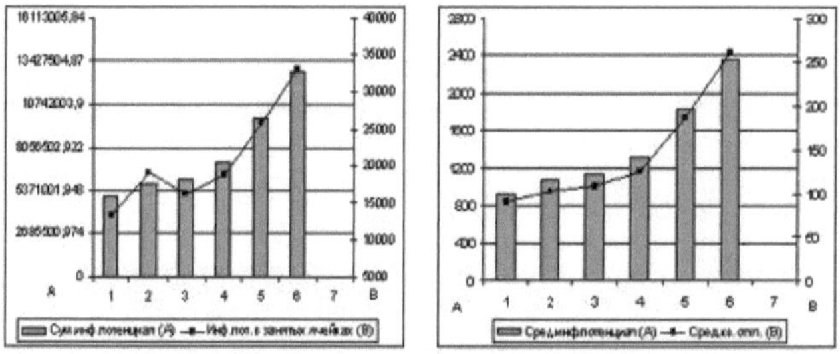

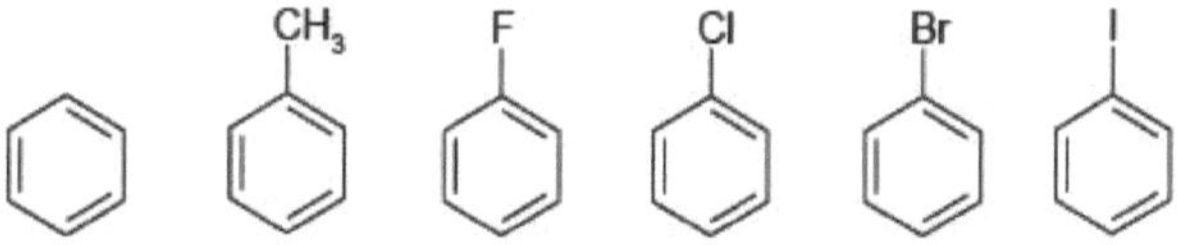

1 2 3 4 5 6

Gradient modulus Divergence

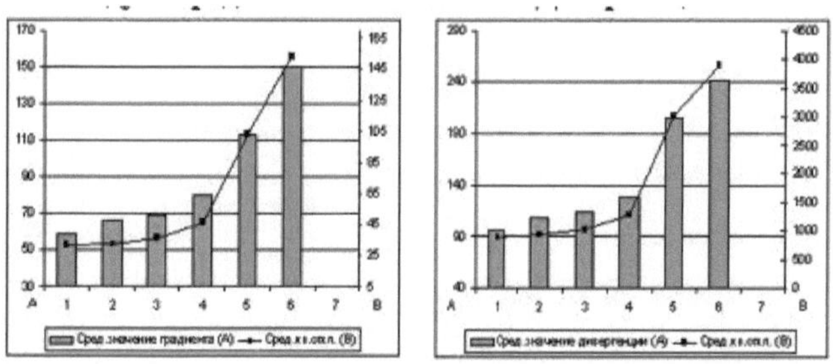

Fig.2.12. Parameters of the scalar information field of monosubstituted benzenes.

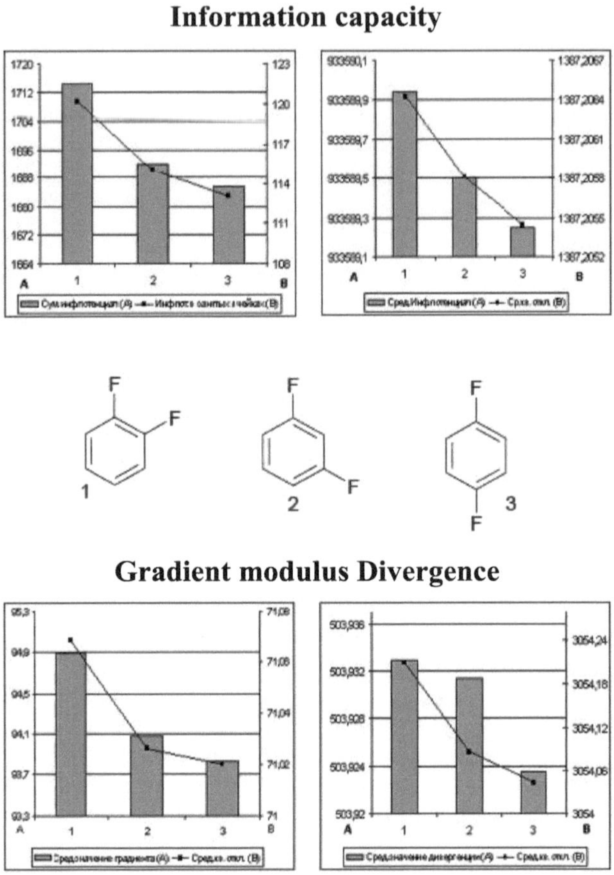

Fig. 2.13. Parameters of the scalar information field of difluorobenzene isomers.

Information capacity

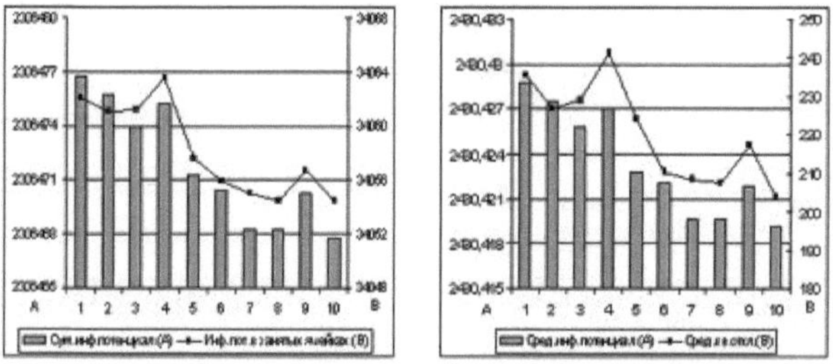

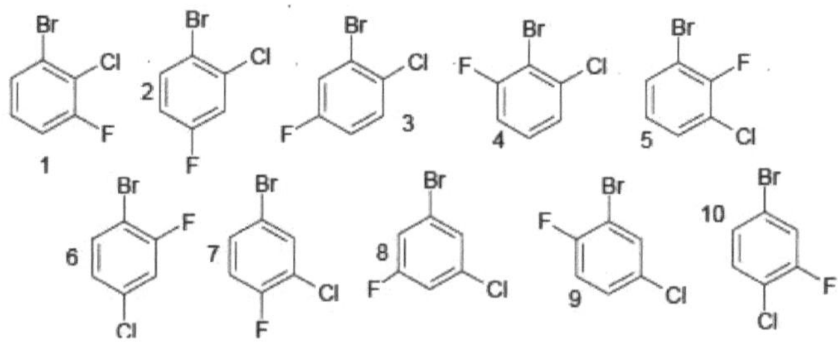

Gradient modulus Divergence

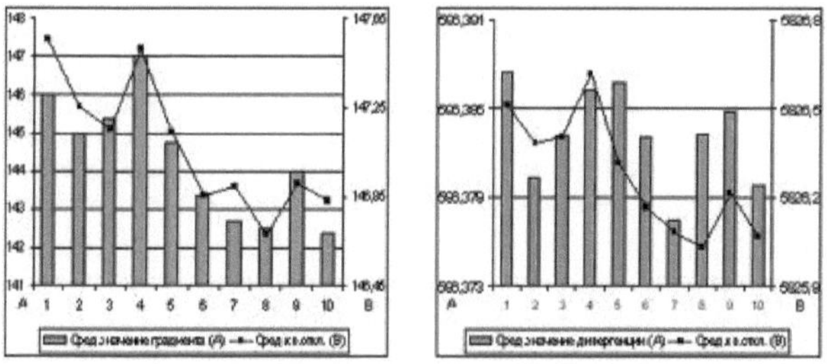

Fig. 2.14. Parameters of the scalar information field of trisubstituted benzenes.

Potential vector IP Divergence of vector IP

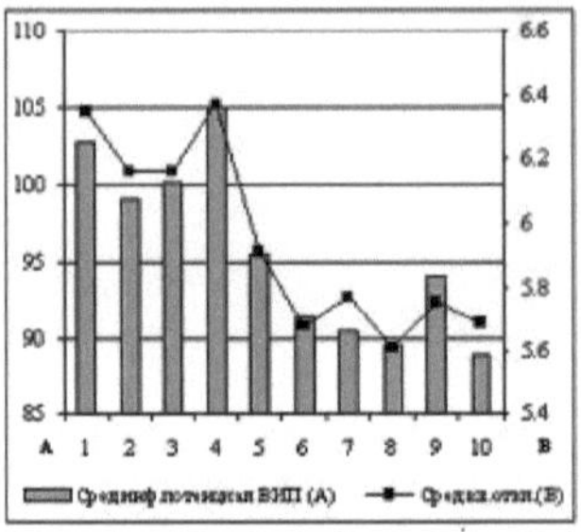

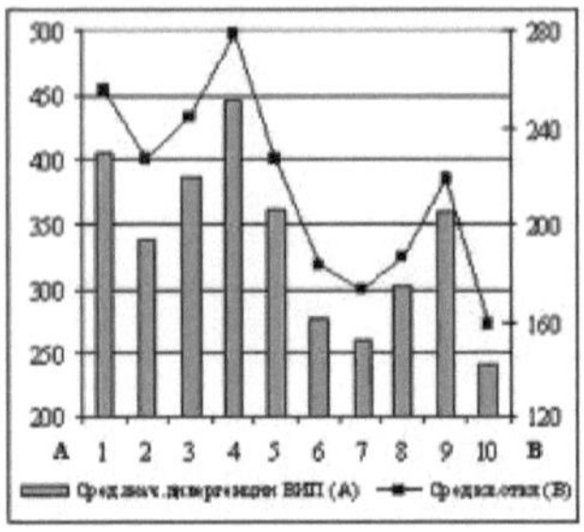

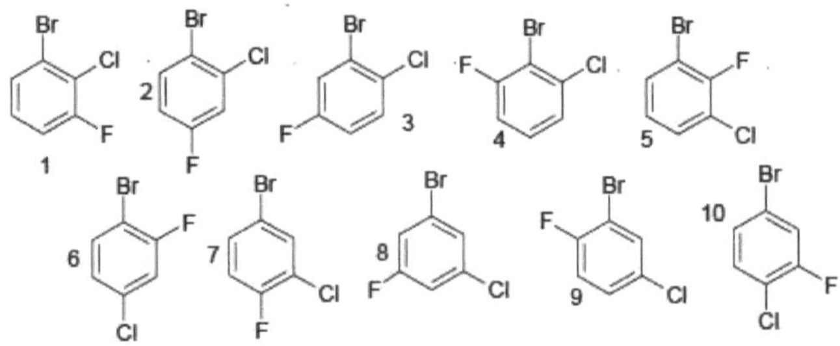

Rotor of the vector IP

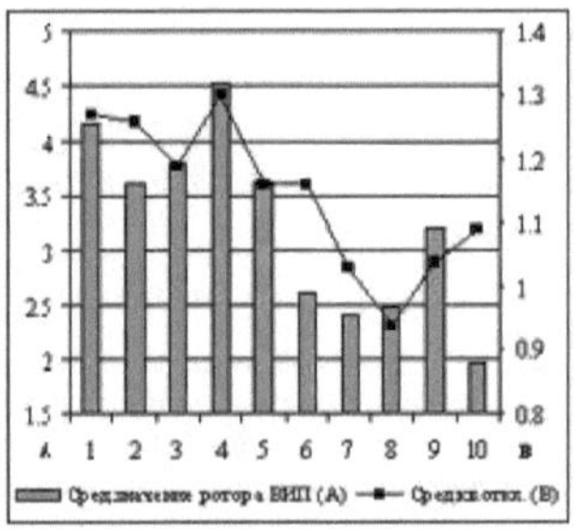

Fig. 2.15. Parameters of the vector information field of trisubstituted benzenes.

CHAPTER 3

FUNDAMENTAL ASPECTS OF THE CONCEPT OF THE INFORMATION FIELD OF A MOLECULE

This section reviews the main ways of developing the developed concept of molecular information field and the resulting applications, in particular, chirality information field (ChIF) modelling, use of information field parameters for analysis of complementary guest-host binding, structural similarity assessment, etc.

3.1. Modelling of the chirality information field

An important application of the developed concept is the possibility of constructing the chirality IP [71]. For the construction of IPC we used a symbiosis of two models: "stereochemical charges" [72, 73] and molecular of the information field [58, 74]. In fact, when constructing the information field of a molecule, we weighted it by "stereochemical charges". In all cases, the weight parameter (dimensionless) was equated to the stereochemical charge or its normalised value:

$$SC_i^{Norm} = \frac{SC_i}{\sum_i^N |SC_i|},$$

where SC_i^{Norm} *is the* normalised value of the "stereochemical" charge for i-th atom; SC_f is the "stereochemical charge" of i-th atom; *i*- the number of atoms in the molecule.

The calculation of "stereochemical charges" is based on the simplex representation of the molecular structure [72]. A molecule is represented as a system of simplexes (four-point fragments) of fixed chirality (R, S, 0 are achiral). For each simplex, a quantitative measure of chirality *(EDF)* is evaluated. The stereochemical charge on an atom is equal to 1/4 of the average *EDF of* all simplexes to which this atom belongs.

In fact, the "stereochemical charge" on an atom reflects the degree of "leftness" or "rightness" of the chiral environment in which that atom is located.

The information potential of the thus constructed chirality field at each point of space reflects the influence of the "left" and "right" fragments (simplexes) of the molecule under study on the dissymmetry of space at this point. In other words, the space is dissymmetrized and becomes "right" near the "right" fragments and "left" near the "left" fragments.

Let us consider the simplest case - a molecule with an asymmetric centre. The IPC of the bromo-fluoro-chloromethane molecule has regions of positive and negative information potential (in other words: "right" and "left" chirality). They do not compensate each other, so, in general, the molecule is chiral and shows "right" influence on the surrounding space (Fig. 3.1). The scheme of partitioning the molecule into simplexes is presented below and the values of "stereochemical charges" for each atom of the molecule are given.

Stereochemical charges

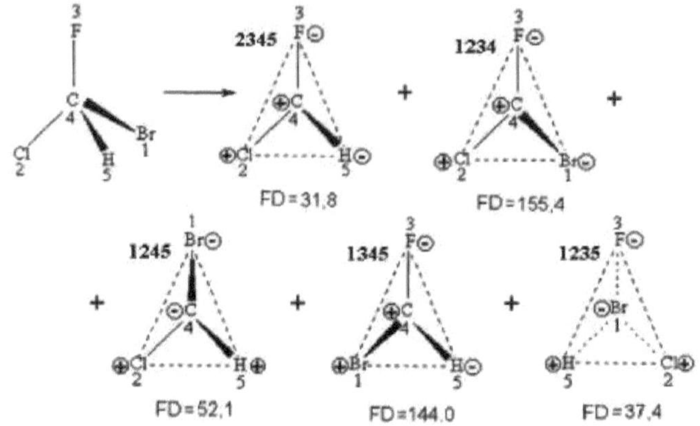

H: -0.08; F: -0.33; C: +0.26; Cl: +0.24; Br: -0.09

If we take as an example the achiral molecule difluoro-chloromethane (symmetry C_s), the picture changes significantly, since the available regions of positive and negative chirality potentials, for obvious reasons, completely compensate each other (Fig. 3.2).

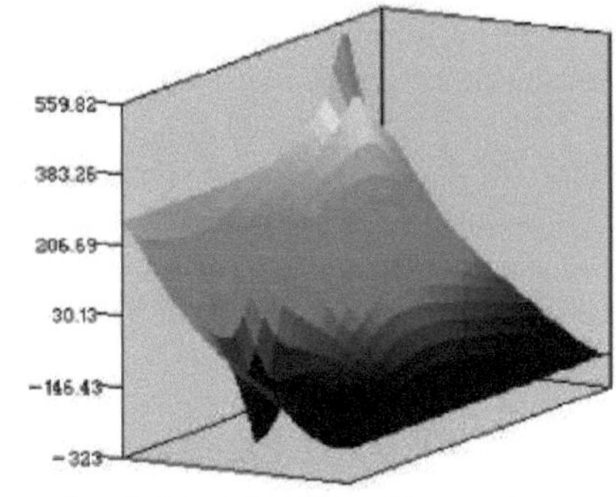

Fig.3.1. Information field of chirality of bromo-fluoro-chloromethane.

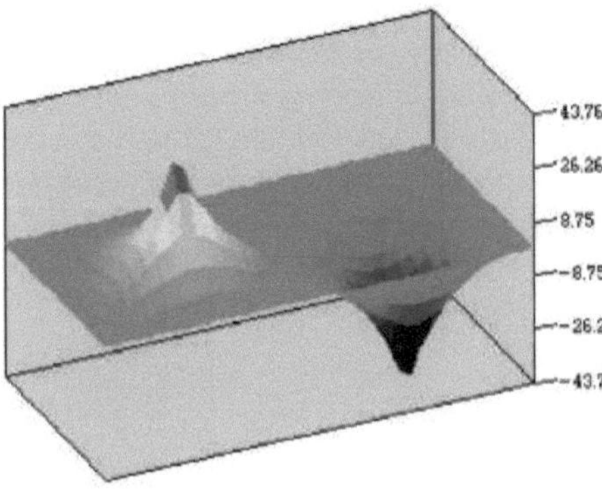

Fig.3.2. Information field of chirality of difluoro-chloromethane molecule.

Another example is the hexagelicene molecule (symmetry C_2). It creates a chirality field of one sign, in this case negative ("left") (Fig. 3.3).

Let us consider another case - the achiral molecule of dibromo-difluoro-dichloroethane. It has two asymmetric centres of opposite configuration (meso-form) and possesses central symmetry C;. If we "cut" this molecule along the main axes of inertia, we can observe that the value of the information potential changes, passing through zero with a change of sign to the opposite one. Moreover, strict symmetry of the picture is observed. And if we summarise the information in the whole observed space, it will correspond to zero chirality: the left part of the molecule compensates the influence of the right part (Fig. 3.4).

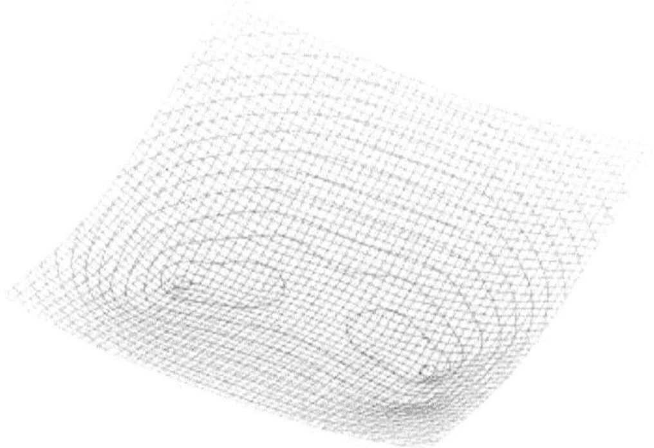

Fig.3.3 Chirality field of hexagelicene molecule.

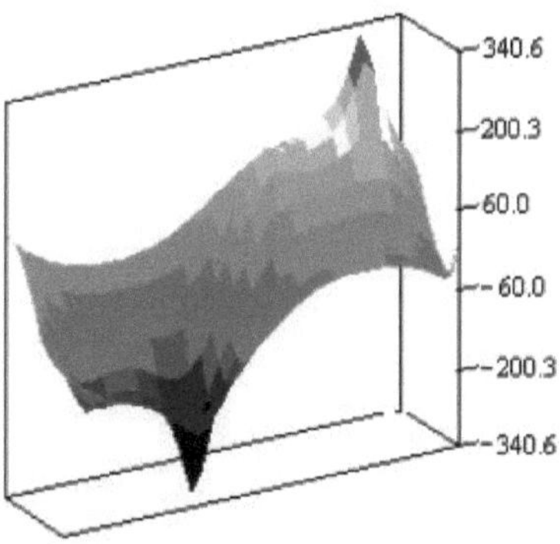

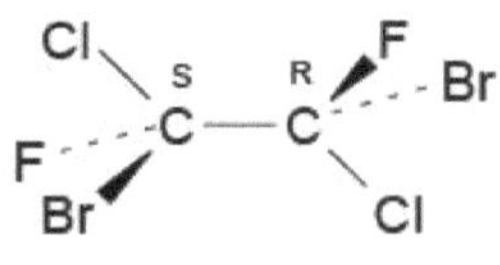

Fig.3.4. Information field of achiral molecule of dibromo-difluoro-dichloroethane.

For the chiral molecule dibromo-difluoro-dichloroethane (symmetry C_2) we naturally do not observe such a picture. No matter how we "cut" the molecule along the main axes of inertia and evaluate the average information potential, it will always be of the same sign; in this case it is positive. That is, this molecule creates only the "right" chiral field around itself, and no other (Fig. 3.5).

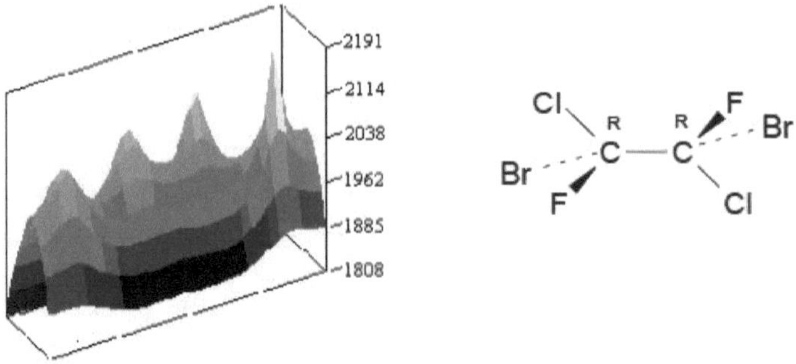

Fig.3.5. Information field of chiral molecule of dibromo-difluoro-dichloroethane.

It is necessary to pay attention to such an interesting fact: in all cases chiral molecules with axial symmetry, create around themselves a field of only one sign. As an illustration, we give the IPC of trans,trans,trans- perhydrotriphenylene. The molecule has symmetry D_3 , has no known elements of chirality (centre, axis, plane), nevertheless it is chiral. This molecule creates a symmetric field of one sign in the surrounding space (Fig. 3.6).

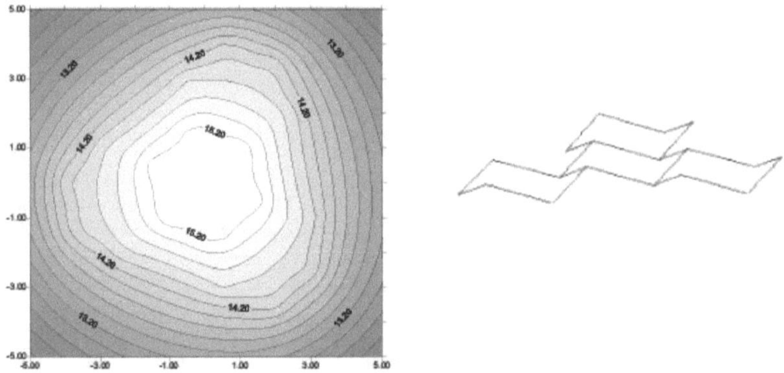

Fig.3.6. Projection of IPC on the RMS plane of the trans,trans,trans-perhydrotriphenylene molecule.

3.2. Analysis of the chirality information field for dissymmetric fullerenes

To investigate the possibilities of IPC, various fullerenes with dissymmetric backbone were analysed, a comparative analysis of IPC inside and outside the molecules was done, and an attempt was made to model a homogeneous IPC.

The following compounds were studied: C32, C60, C76, C78. The IPCs were analysed for these compounds.

Let us consider the chirality information field of fullerene C76. The molecule has symmetry D_2 . There is no known chirality element here [75]. This molecule creates a chirality field of only one sign in the surrounding space.

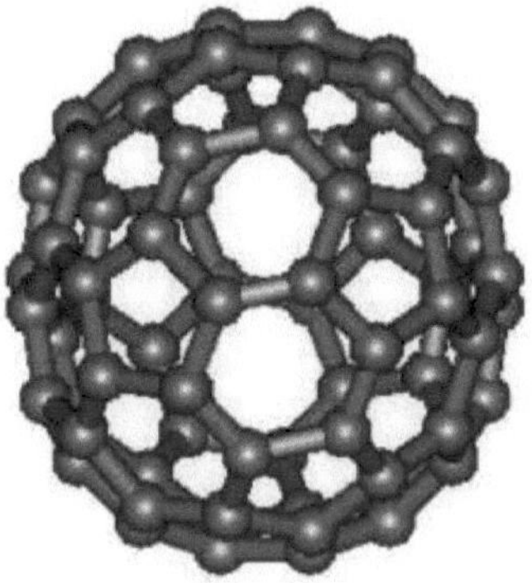

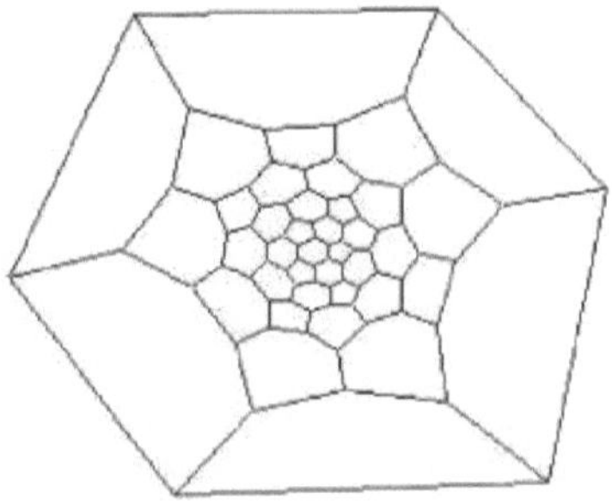

It is interesting to note that the qualitative picture for the information fields inside and outside the molecule is almost identical, but when considering smaller molecules C32 and C60, the field inside the cavity of the molecule is, as one would expect, more intense.
It should be noted that the IPC type reflects the peculiarities of the molecule geometry, determining the position of symmetry elements. The above is well illustrated by the example of the chirality information field of fullerene C76, where the directions of the second-order symmetry axes are clearly expressed (Fig. 3.7)

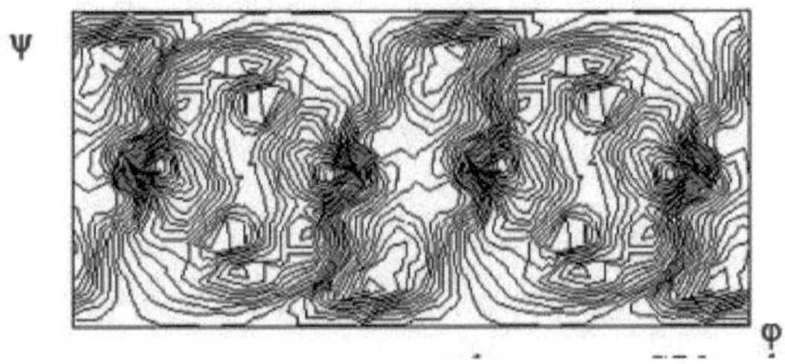

IP chirality outside the fullerene C76 molecule in spherical coordinates.

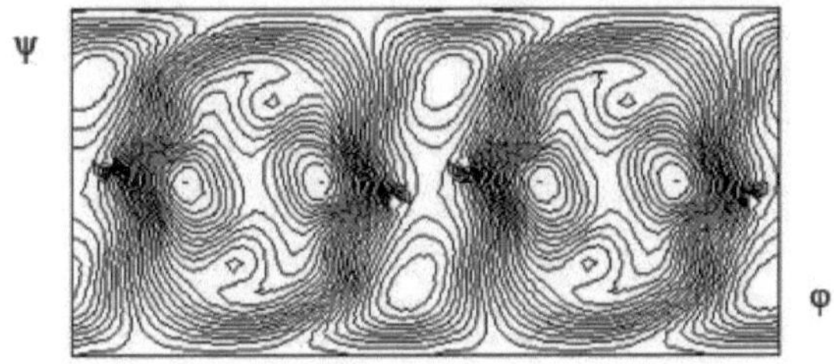

IP chirality within a fullerene C76 molecule in spherical coordinates.

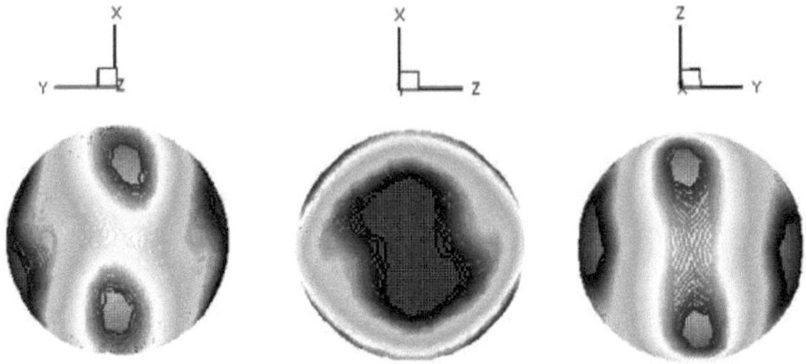

Fig.3.7. IPC on a spherical surface around the fullerene molecule C76: the direction of symmetry axes C_2 corresponds to the coordinate axes.

In all cases, the signs of the chirality field are determined by the type of enantiomer ("left" or "right"). Naturally, always a pair of enantiomers generates chirality fields of opposite signs.

IP chirality maps of the chirality inside and outside the molecules for C32, C60 and C78 fullerenes are presented below.

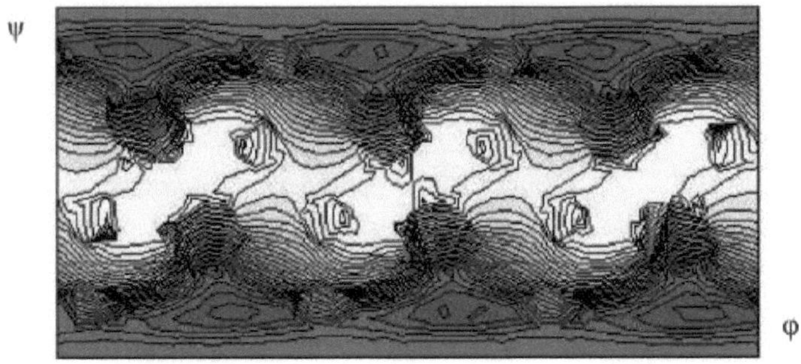

IP chirality inside the fullerene C32 molecule in spherical coordinates

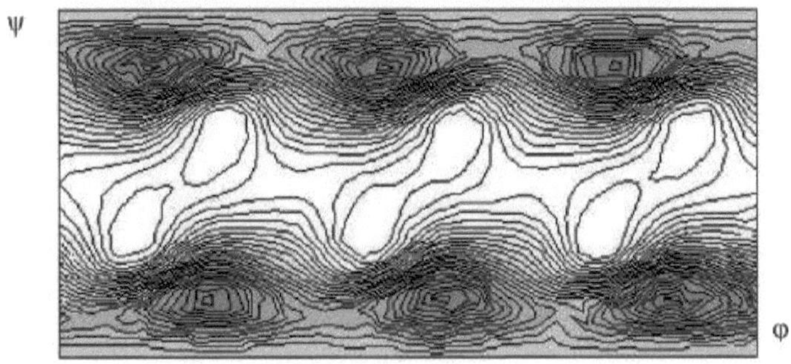

IP chirality outside the fullerene C32 molecule in spherical coordinates

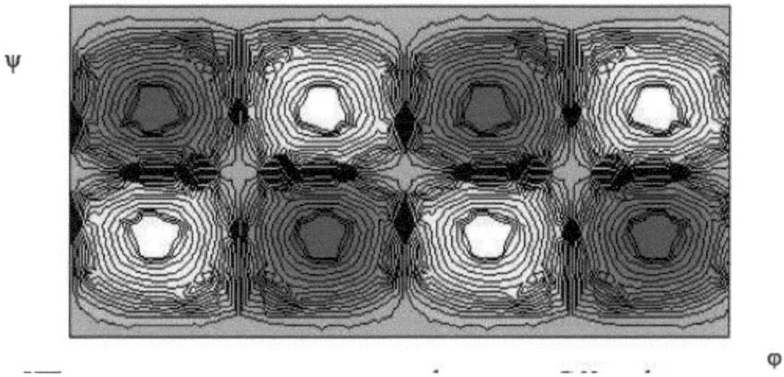

IP chirality inside the fullerene C60 molecule in spherical coordinates

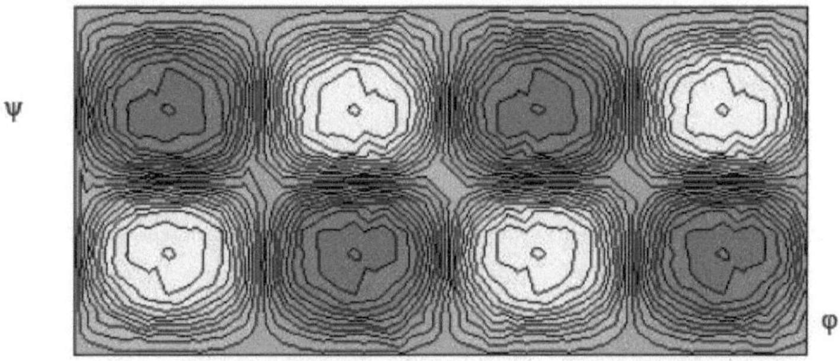

IP chirality outside the fullerene C60 molecule in spherical coordinates

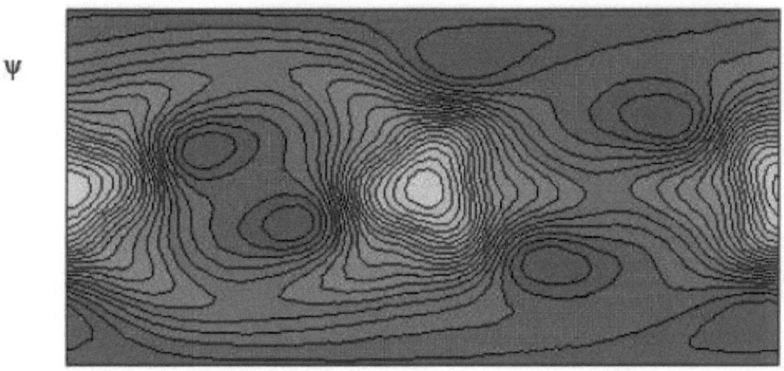

IP chirality inside the fullerene C78 molecule in spherical coordinates

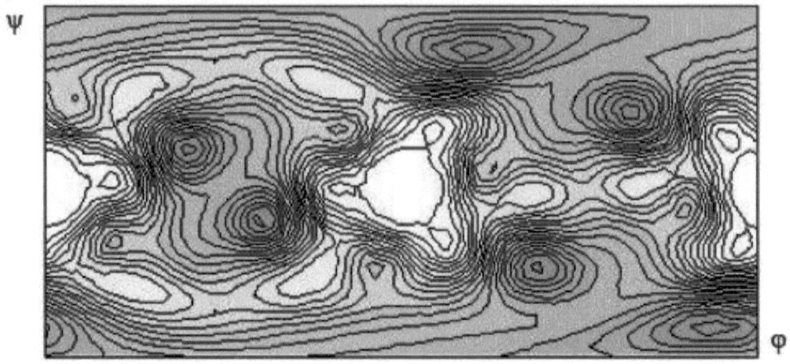

IP chirality outside the fullerene C78 molecule in spherical coordinates

For C76, the differences are smaller than for C32. For C60, as one would expect, the field is achiral in general, although there are "left" and "right" regions, but they compensate each other.

3.3. Modelling of homogeneous chirality IP

Within the framework of the developed chirality FE model, an attempt was made to model a homogeneous chirality FE. For this purpose, a model system of a ring coil made of "wire" wound on a torus-shaped core was compiled. The investigated model system is a sequence of identical atoms located on the turns of the modelled toroid (Fig. 3.8).

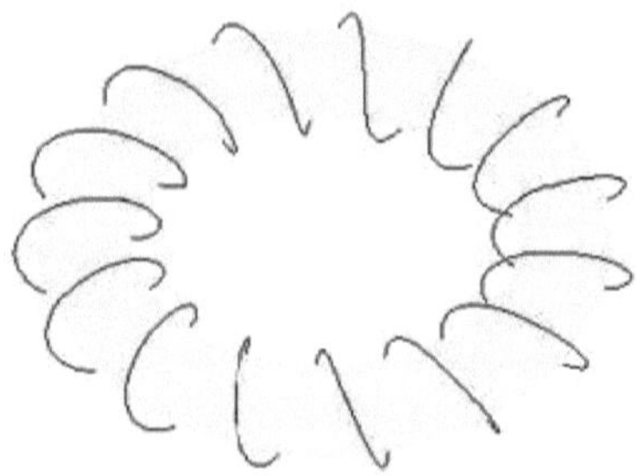

Figure 3.8. Ring coil (toroid)

Analysis of the chirality IP of this system showed that a homogeneous field is formed inside such a toroid, while at a distance from the toroid, as one would expect, there is a decay of the chirality IP (Fig. 3.9).

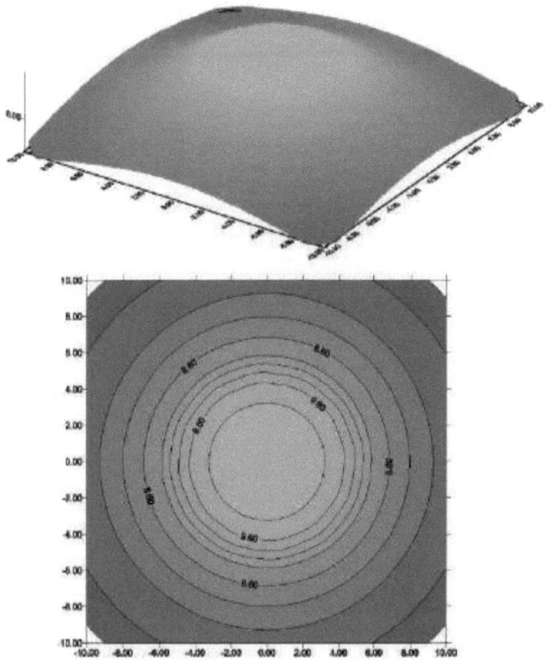

Figure 3.9. Homogeneous chirality IP.

If a real molecular system of this type can be created, it may have interesting stereochemical features.

Thus, the proposed model allows us to describe the ability of chiral molecules to dissymmetrize their surrounding space and, in all probability, the parameters of the chirality information field can be used as descriptors for solving various structure-property problems.

3.4. Analysis of molecular recognition in guest-host complexes within the concept of molecular information field

The possibilities of the developed concept of the IP of molecules can be used in analysing the efficiency of complementary guest-host binding [76, 77]. As follows from the previous sections, the information field of a molecule actually characterises the "structuredness" of the surrounding space. The influence of a molecule on the surrounding space is most clearly reflected by the divergence of the information field. It seems to us that a comparative analysis of the divergence of the fields of the guest and host IP gradients will allow us to quantitatively assess the efficiency of their complementary interaction.

Two approaches were used to solve such a problem.

1. The measure of similarity of two information fields can be a value that is calculated as a correlation coefficient between the information parameters of all corresponding pairs of

points that represent the corresponding information fields (Fig. 3.10):

$$r = \frac{n\sum_{i=1}^{n} h_i g_i - \sum_{i=1}^{n} h_i \sum_{i=1}^{n} g_i}{\sqrt{n\sum_i g_i^2 - \left(\sum_i g_i\right)^2} \cdot \sqrt{n\sum_i h_i^2 - \left(\sum_i h_i\right)^2}}$$

where *hi* and *e, are the* information parameters (potential or divergence of the gradient field) of the host and guest, respectively, at point *i*;

n is the number of points in the investigated region of space;

i is the number of the point.

The closer *r is* to 1, the more "similar" the corresponding fields are.

Note that in this case the result depends on how "successfully" the guest g_i and host *h* molecules are combined.

2. Only points on the plane that is the mean-square plane of the molecule are considered. The specified points are approximated by a spiral line (e.g., "right") (Fig. 3.11).

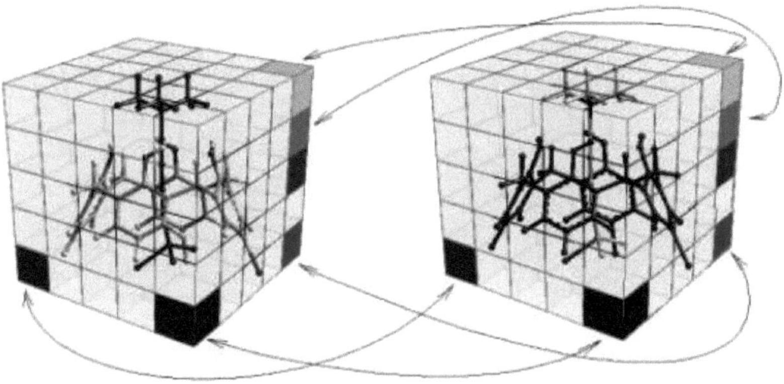

Fig.3.10. Comparison of information parameters in given points of space.

At these points, the parameters of the information fields (potential or gradient field divergence) of the host (h_i) and guest (g_i) molecules, respectively, are calculated. For points ordered relative to the spiral line, autocorrelation functions are calculated based on the parameters of the information fields

separately for the host and guest molecule:

$$R_h(t) = \frac{\sum_{i=1}^{n}\left(h_i - \bar{h}\right)\cdot\left(h_{i+t} - \bar{h}\right)}{n \cdot S_h^2}; S_h = \sqrt{\left[\sum_{i=1}^{n}\left(h_i - \bar{h}\right)^2\right] / n}; \bar{h} = \frac{\sum_{i=1}^{n} h_i}{n}$$

$$R_g(t) = \frac{\sum_{i=1}^{n}\left(g_i - \bar{g}\right)\cdot\left(g_{i+t} - \bar{g}\right)}{n \cdot S_g^2}; S_g = \sqrt{\left[\sum_{i=1}^{n}\left(g_i - \bar{g}\right)^2\right] / n}; \bar{g} = \frac{\sum_{i=1}^{n} g_i}{n}$$

where $t = 1, 2, 3 \ldots 60$.
The measure of field similarity is the % value, which is defined as the Euclidean metric of the corresponding FE parameters:

$$\chi = \sqrt{\sum_{t=1} \left[R_h(t) - R_g(t)\right]^2}$$

The smaller the % value, the more "similar" the corresponding fields are.
Note that in this approach it is not necessary to precisely match the guest and host molecules, it is sufficient to determine the mutual orientation of the molecules in the complex.

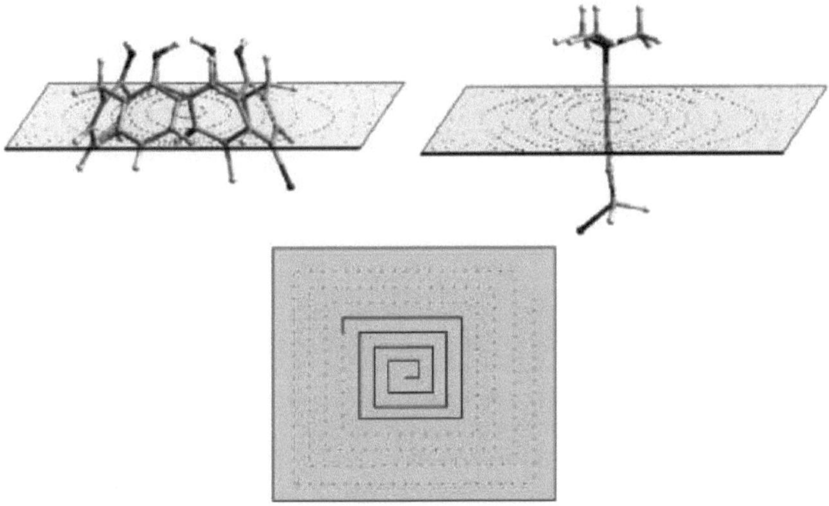

Fig.3.11. Projections of IP on the RMS planes of the compared molecules.
The following model examples were used to analyse complementary guest-host binding: host molecules are enantiomeric (R_h and S_h) calixarenes:

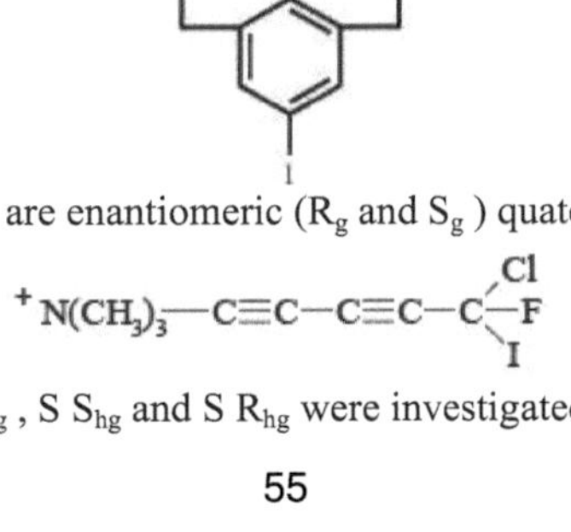

and guest (guest) molecules are enantiomeric (R_g and S_g) quaternary ammonium salts:

The complexes R R_{hg} , R S_{hg} , S S_{hg} and S R_{hg} were investigated.

At the initial stage of the work, these structures were optimised using a semi-empirical quantum mechanical method (MNDO, HyperChem [78]). Analysing the obtained energies of the complexes, it can be noted that the complexes R S_{hg} and S R_{hg} are energetically more favourable (Table 3.1). Further, the mass distribution fields and chirality IPs were analysed for each complex. The complementarity of the guest-host interaction was assessed by comparing the information fields of the corresponding structures using the methods proposed above.

The study of all diastereomeric complexes R R_{hg} , R S_{hg} , S S_{hg} , S R_{hg} showed that the highest complementarity is achieved in the complexes R S_{hg} or S R_{hg} . The obtained results are in agreement with the estimation of stability of the corresponding complexes on the basis of semiempirical quantum-chemical calculations, i.e. for the most stable complexes the complementarity of the corresponding fields is maximal (Table 3.1).

Table 3.1.

Results of complementary binding analysis of host and guest molecules.

complex	E_{MNDO}	$r\ div\_mass$	rdiv chiral	rpot_mass	rpotchiral	Xdivmass	Xdivchiral
RhR_g	-7983.57	0.037	-0.057	0.753	-0.362	2.26	4.99
RhSg	-7984.92	0.048	-0.078	0.820	-0.510	1.85	4.68
ShSg	-7983.58	0.039	-0.055	0.748	-0.361	2.23	4.95
ShRg	-7984.89	0.047	-0.075	0.816	-0.508	1.87	4.63

E_{MNDO} is the energy of the host-guest complex estimated by the semi-empirical MNDO method;

$rd\ _{\_ivmass}$ is the correlation coefficient between the divergence parameter sets of the IP gradient fields weighted by atomic masses for host and guest molecules;

$r_{div\ chira}$ *is the* correlation coefficient between the sets of divergence parameters of the IP gradient fields weighted by "stereochemical charges" for host and guest molecules; $r_{pot\ mass}$ *is the* correlation coefficient between the sets of information field potential parameters weighted by atomic masses for host and guest molecules; $r\ \_\ h_{potcira}$ *i is the* correlation coefficient between the sets of information field potential parameters weighted by "stereochemical charges" for host and guest molecules;

$Xd\__{ivmass}$ *is a* mass-weighted similarity measure of information fields;

$Xdiv_{\ chiral}$ *is a* measure of similarity of information fields weighted by "stereochemical charges".

Maps of the corresponding information fields of atomic mass distributions and "stereochemical charges" are presented in Appendix A.

It is suggested that the developed approach may be promising for the molecular design of host molecules capable of complementary binding of specific guest molecules. Naturally, such an approach is applicable in cases where differences in the stability of guest-host complexes are due to differences in the degree of complementarity.

3.5 Quantification of isosterism (bioisosterism) of organic compounds

Nowadays, the process of drug development starts in most cases with the search for a leader compound - a structural prototype of a future drug that has the desired pharmacological effect. Since the leader compound, as a rule, also causes undesirable side effects or has poor pharmacokinetic characteristics, there is a need to optimise the leader compound, consisting in synthetic modification of the structure of the leader compound in order to increase its activity, reduce toxicity and improve selectivity of action. This often involves approaches involving the synthesis of structural analogues of the leader compound. One such approach used to create more effective and safer drugs is bioisosteric substitution [79].

A bioisoster is a compound obtained by replacing one atom (group of atoms) with another atom (group of atoms) that retains the biological activity of the parent compound. As a result of such replacement, new substances with improved properties compared to the parent compound can be obtained.

Bioisosters are divided into two groups: classical isosters, having the same size, shape and configuration of the outer electron layer, and non-classical bioisosters, i.e. atoms or groups not possessing the characteristics of classical isosters, but having the same biological activity.

Classical isosteric substitution involves the substitution of like for like, taking into account the number of atoms, valence, degree of unsaturation and aromaticity, and becomes bioisosteric if the biological effect of the compound is also preserved.

When analysing non-classical bioisosters, the question arises as to why compounds with completely different chemical formulas can exhibit the same biological activity. The search for an answer to this question led to the development of the concept of bioisosterism in the works of K. Ganch [80]. In his work, K. Ganch did not search for similarities in the chemical formulas of non-classical bioisosters, but tried to explain the reason for the same biological action of the compounds under study using the mathematical approach he had proposed earlier, describing structure-activity relationships in terms of three main differences in the properties of physiologically active compounds: steric, electronic, and hydrophobic. This work concluded that, in general, for compounds to have similar biological effects, they must have similar hydrophobic, electronic and steric characteristics. However, in many cases, it is not necessary for each of these properties to be the same for several substances to have the same effect on a biological system. The biological response may, for example, be determined only by the hydrophobic characteristics of a substance, while electronic and steric characteristics do not play a significant role in this effect and vice versa. That is, for non-classical bioisostereomers, biological activity may be determined by a single (or a group of properties) property that has the same significance for a series of given compounds, while other properties have no influence on the manifestation of this type of activity.

Using the concept of molecular information field proposed by us, we attempted to quantitatively assess the bioisosterism of various compounds [81]. Two groups of compounds were analysed, one of which includes representatives of classical bioisosters (isosters) and the other group includes non-classical bioisosters. As an example, we considered the compounds given in [79]: Classical bioisosters

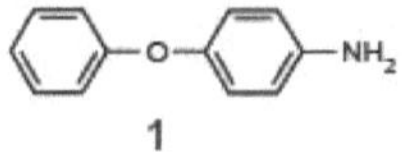
1

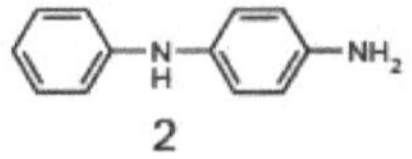
2

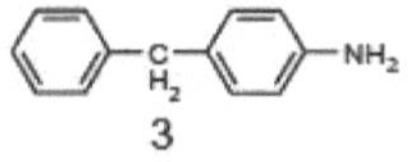
3

Non-classical bioisosters

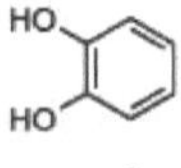
4

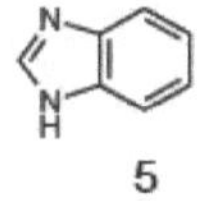
5

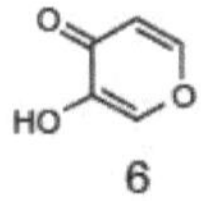
6

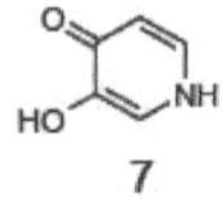
7

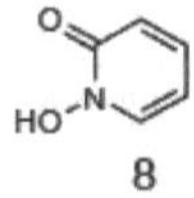
8

The information fields of these molecules were constructed, weighted by various properties of atoms: masses, van der Waals radii, lipophilicity, and charges. It should be noted that the divergence parameter of the information field of molecules was calculated because, as noted earlier, this parameter more clearly reflects changes in the molecular structure.

After that, the similarity of the constructed information fields of the investigated molecules was quantitatively evaluated. The similarity measure of two information fields is calculated as the correlation coefficient between the corresponding sets of information parameters (pairs of points), which represent the corresponding information fields:

$$r = \frac{n\sum_{i=1}^{n} h_i g_i - \sum_{i=1}^{n} h_i \sum_{i=1}^{n} g_i}{\sqrt{n\sum_i g_i^2 - \left(\sum_i g_i\right)^2} \cdot \sqrt{n\sum_i h_i^2 - \left(\sum_i h_i\right)^2}}$$

where h_i and g_i *are* information parameters (divergence), respectively, of the first and second molecules at point *i*; *n is the* number of points in the investigated region of space; *i is the* point number.

The closer the value of *r is* to 1, the more "similar" the corresponding fields are.

When analysing the different information fields of classical bioisosters (Fig. 3.12) and non-classical bioisosters (Fig. 3.13), it can be seen that qualitatively the information fields of classical bioisosters are more similar compared to non-classical bioisosters.

Quantitatively, the similarity of the molecules under study can be viewed as the distance between the points corresponding to each molecule in the information space.

As a result of analysing the information fields of classical bioisosters, it was found that their respective information fields are quite similar and the distance between the corresponding pairs of points is almost the same. As can be seen from Fig. 3.14, for classical bioisosters, apart from the fact that the similarity measure varies equally in different cases, the distance spread for them is smaller than for non-classical bioisosters, for which the similarity measure varies differently in each case. In Fig. 3.14 we used *S,* a conditional distance scale ($S_i\ j = \backslash S_i\ -Sj\backslash$), based on which the structural similarity of molecules *i* and *j is* estimated. The scale was obtained as a result of the operation of "multidimensional scaling" of Rij [82].

Thus, evaluating quantitatively and qualitatively the similarity of IPs of classical bioisosters, we can conclude that molecules 1, 2, 3 have close structures, properties and corresponding IPs and their biological action (immunological activity) follows from their structural similarity.

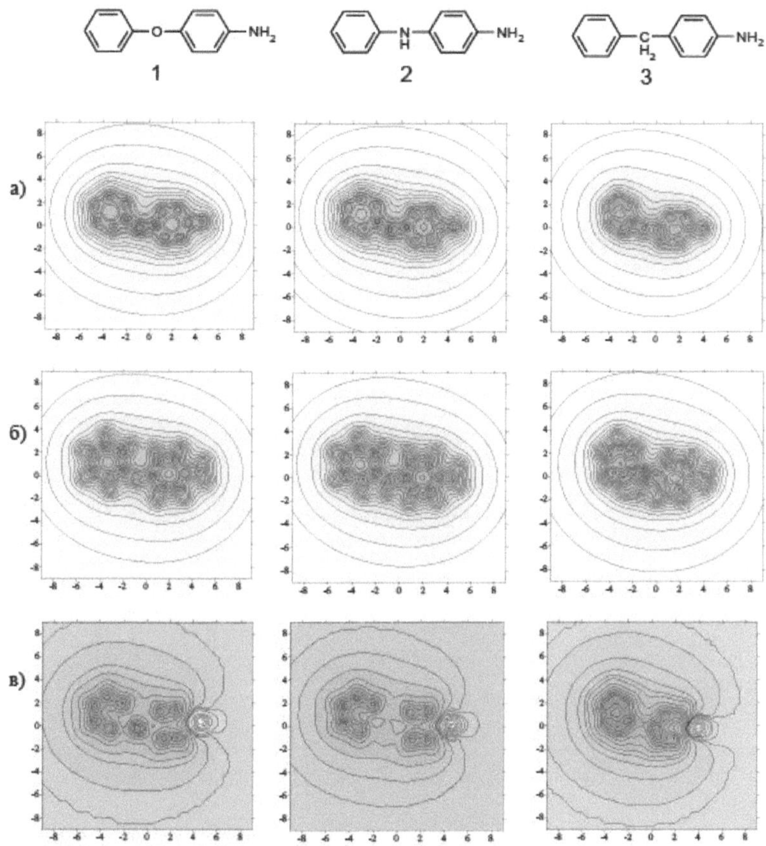

Fig. 3.12. Information fields of classical (1-3) bioisosters weighted by the following properties: a) atomic mass, b) van der Waals radius, c) lipophilicity.

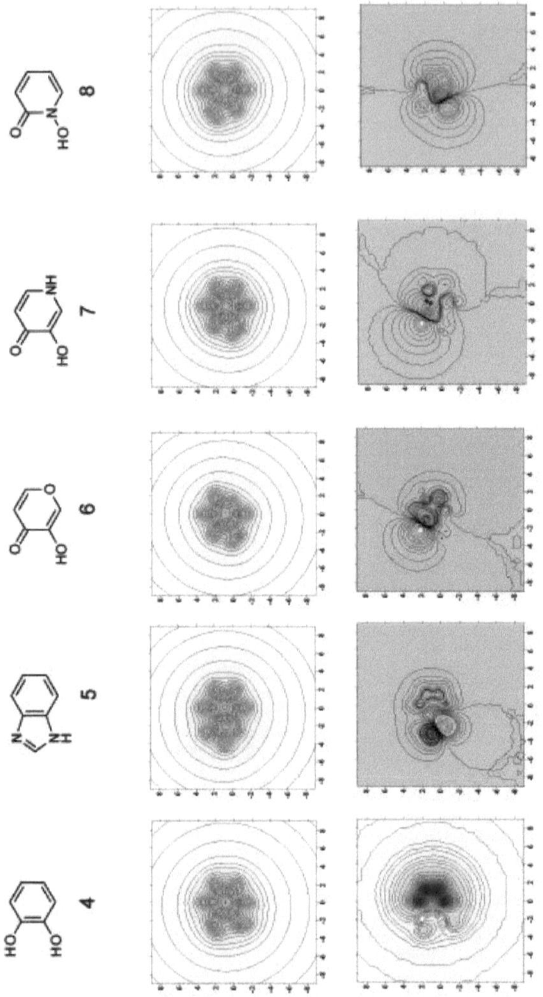

Fig. 3.13 Information fields of non-classical (4 - 8) bioisosters weighted by the following properties: a) van der Waals radius, b) lipophilicity.

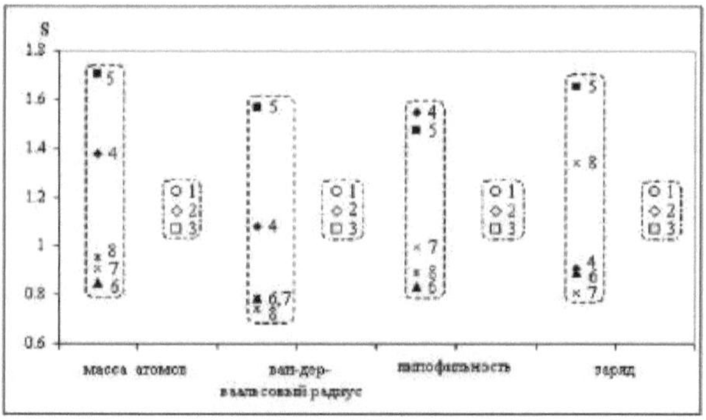

Fig.3.14. Assessment of structural similarity / difference of classical (1-3) and

non-classical (4 - 8) bioisosters.

As a result of analysing the information fields of non-classical bioisosters weighted by different properties, it was found that the corresponding IPs for molecules 4, 5, 6, 7, 8 are quite different, the distance between the corresponding pairs of points being significantly different. That is, for non-classical bioisosters, the measure of structural similarity depends on structural characteristics (atomic mass, van der Waals radii, lipophilicity, etc.)

3.6 Calculation of information parameters of essential amino acids

In the present work, a systematic analysis of the characteristics of the information fields of all essential amino acids was carried out (Fig. 3.15, Fig. 3.16). A total conformational analysis was previously performed for these compounds. A conformational search [78] was carried out, as a result of which a set of the most stable conformations within the energy gap of 2 kcal/mol was determined for each amino acid. In Fig. 3.15 and Fig. 3.16 show the most favourable conformers of amino acids, their fraction in conformational equilibrium (P) and essentiality (A) according to Wolkenstein [44]. The conditional values of amino acids (indispensability parameters) were proposed by M.V. Wolkenstein on the basis of the following hypothesis. The known data on protein structures made it possible to analyse the interchangeability of amino acid residues in protein molecules. On its basis, a scale of values of amino acid residues according to the degree of their indispensability was proposed. In other words, the less frequently a residue is substituted, the more valuable it is. Only those residue substitutions that are caused by single substitutions in the corresponding codons are considered. Data relating to jointly degenerate codons are summarised and averaged.

For all selected conformers, FEs weighted by different properties of atoms (mass, refraction, electronegativity) were modelled [83]. The construction area of the IPs was varied. The IP parameters were calculated in spherical regions around molecules with radii ranging from 4.5 A (the minimum region of space where each of the studied amino acids fits entirely) to 21A. In all cases, at each indicated point, the potential, antigradient and divergence of the IP gradient field were determined according to the formulae given in Section 2.3. Further, the information characteristics of each molecule were determined as a weighted average of the corresponding parameters of its conformers. The weighting coefficients are equivalent to the probabilities of realisation of conformers.

The following were used as structural characteristics of molecules: minimum and maximum values, values of arithmetic mean and standard deviation (SD) for all types of considered FE characteristics.

Thus, taking into account the variation of the sizes of the regions of space and the different properties of the atoms on which the IPs were weighted, 432 structural characteristics were analysed for each molecule. All these characteristics were further compared with the parameters of amino acid essentiality.

The closest pairwise (IP parameters - AK indispensability) nonlinear correlations (Lorentz function) are observed for the RMS of IP divergence. As can be seen from Fig. 3.17, the best correlations with irreplaceability are the indicated mass-weighted IP parameters. A certain tendency of the influence of the size of the investigated space around the molecule on the tightness of the correlation can be seen.

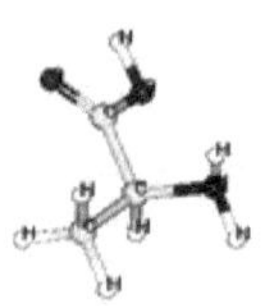

аланин (1)
P= 0,11; A=0,52

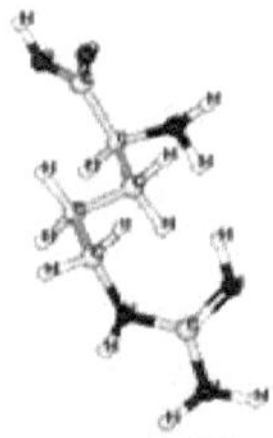

аргинин (2)
P=0,05; A=0,6

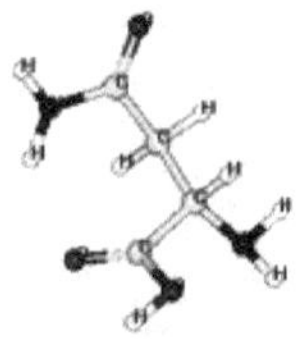

аспарагин (3)
P=0,20; A=0,79

аспарагиновая кислота (4)
P=0,06; A=0,77

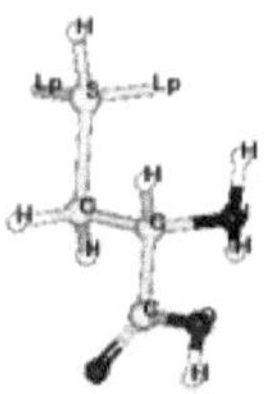

цистеин (5)
P= 0,05; A=1,12

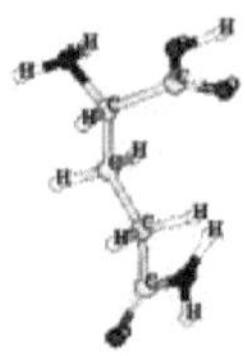

глутамин (6)
P= 0,05; A=0,86

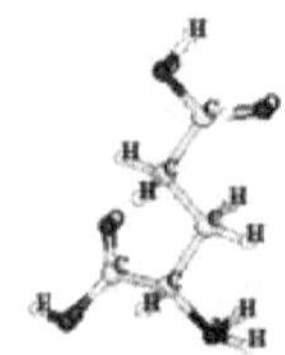

глутаминовая кислота (7)
P=0,05; A=0,76

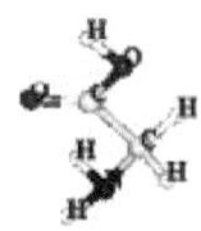

глицин (8)
P= 0,13; A=0,56

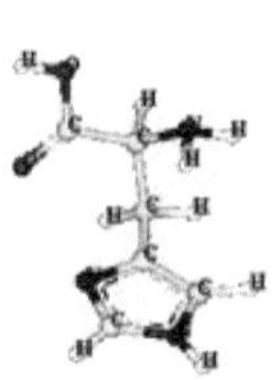

гистидин (9)
P= 0,07; A=0,94

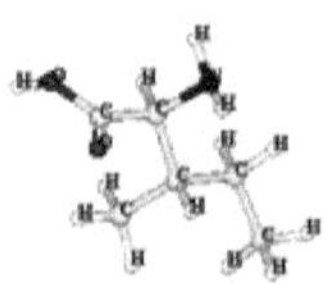

изолейцин (10)
P= 0,07; A=0,65

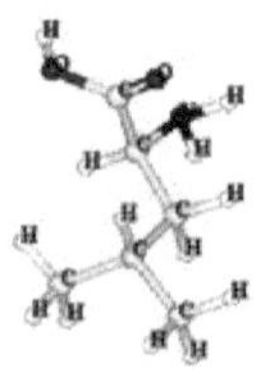

лейцин (11)
P= 0,06; A= 0,58

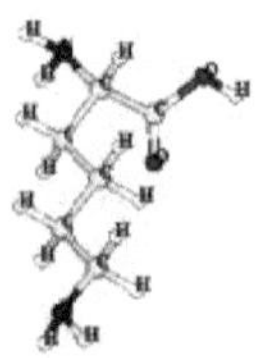

лизин (12)
P= 0,05; A=0,81

alanine (1) arginine (2) asparagine (3) asparagic acid(4) cysteine (5) glutamine (6) glutamic acid(7) glycine (8) histidine (9) isoleucine (10) leucine (11) lysine (12)

Fig. 3.15. The most stable conformers of essential amino acids (molecules 1 - 12) (P - conformer fraction in conformational equilibrium, A - indication of essentiality of AA).

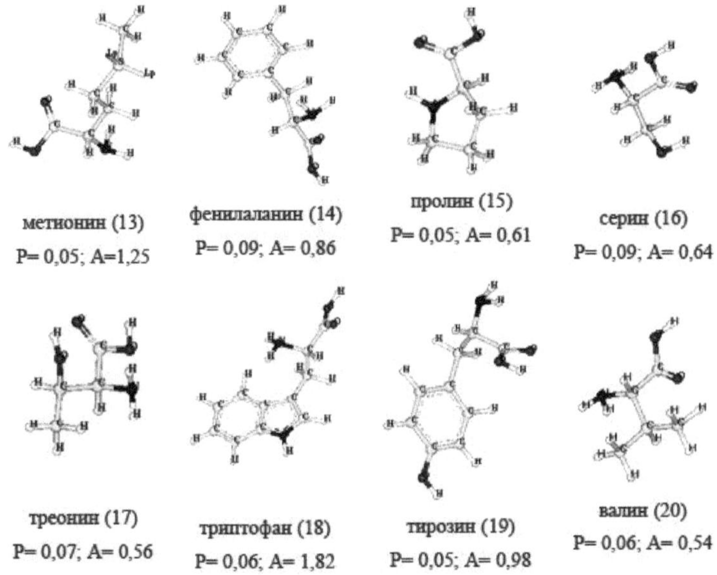

methionine (13) phenylalanine (14) proline (15) serine (16) threonine (17) tryptophan (18) tyrosine (19) valine (20)

Fig. 3.16. The most stable conformers of essential amino acids (molecules of 13 - 20) (P is the conformer fraction in conformational equilibrium, A is the indication of the essentiality of AK).

The maximum correlations are observed for the RMS of the IP divergence calculated for a spherical region around a molecule with a radius of 17 A or more. The best "FE characteristic - indispensability" correlation is shown in Fig. 3.18.

R (coefficient of cor)

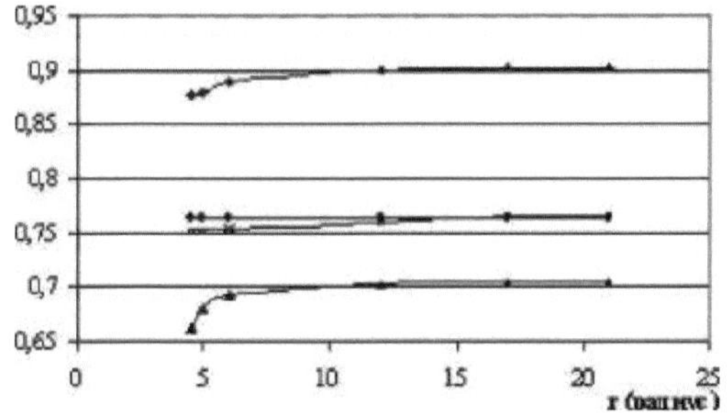

-♦-mass, --re fr. -zpektrootr. -zar.

Fig. 3.17. Dependence of the correlation coefficient of amino acids essentiality with their IP parameters on the size of the region of the investigated space and characteristics of atoms used as weight parameters.

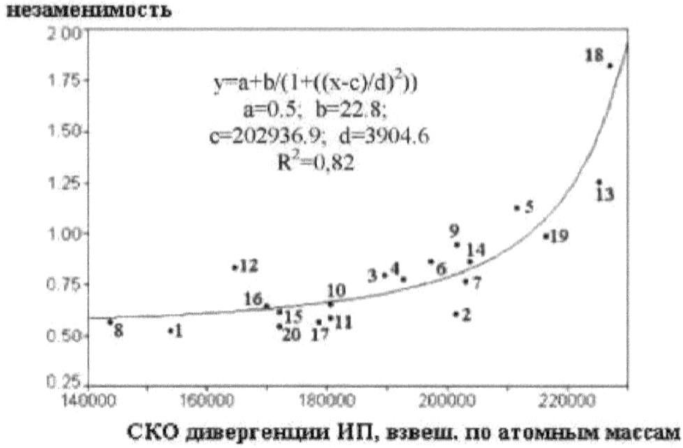

Fig. 3.18. Dependence "RMS of divergence of mass-weighted IPs - essentiality (A)" (the numbers near the graph points correspond to the numbers of amino acids (Fig. 3.15 and Fig. 3.16)).

If we consider the information content of these molecules [31], it does not correlate with the essentiality factor (correlation coefficient = 0.198) (Fig. 3.19).

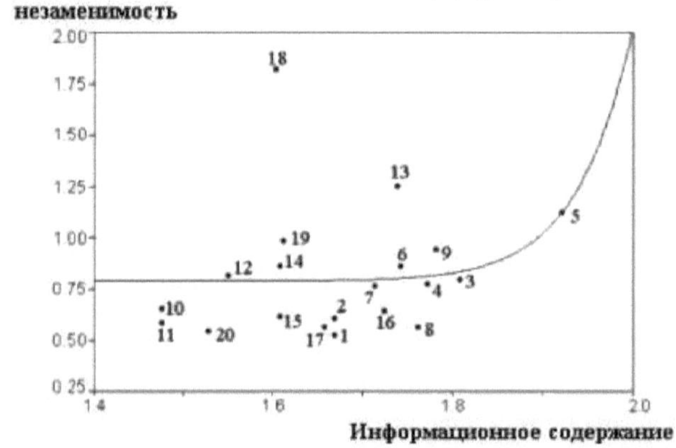

Fig. 3.19. Dependence "Information content of amino acids - non-substitutability (A)" (numbers near the graph points correspond to the numbers of amino acids (Fig. 3.15.)).

A study of the information fields of amino acids found that:

1) The divergence of the information field best reflects the peculiarities of the molecule structure.

to a certain extent the degree of influence of the molecule structure on the surrounding space. In other words, the higher the RMS of IP divergence, the stronger the molecule structures the surrounding space, i.e., transmits information about its structure.

2) There is a rather close connection between the informational influence of an amino acid on the environment and its indispensability. This allows us to hypothesise that at the

early stages of abiogenesis, hereditary information was transmitted without the participation of nucleic acids (which did not exist at that time) [84]. That is, new polypeptides (oligopeptides) were formed directly on the matrix from already prepared peptide molecules. Thus, the assumption seems to be quite justified: the more essential the informational role of an amino acid (as an element of a polypeptide), the more indispensable it is.

3.6. Analysis of mutual correlations of FE parameters

When considering different FEs, we have often encountered the fact that the information parameters in different points of the investigated space are correlated with each other. In this connection, we analysed the mutual correlations of the IP parameters in the cells of the molecular lattice. A series of amino acids and a series of steroids were considered (Appendix B). The molecules in the series were superimposed on each other and placed in a cubic lattice (Figure 3.20). In each lattice cell, FE parameters were calculated and then mutual correlations were analysed.

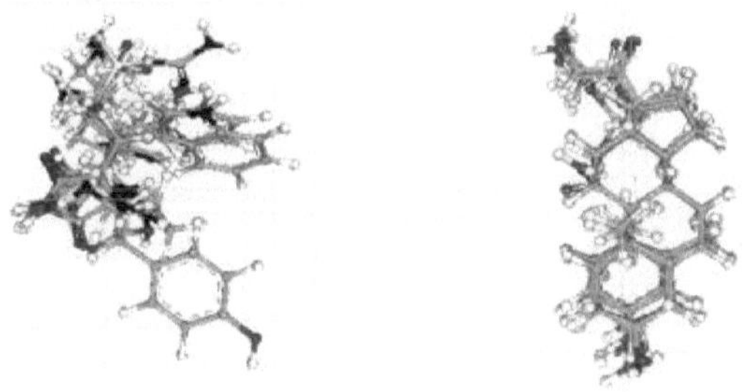

a) б)

Fig. 3.20. Superimposed molecules of series of: a) amino acids and b) steroids in molecular lattice cells.

The mean correlation coefficient for the steroid series (cor. coefficient = 0.83) was found to be higher than that for the amino acid series (cor. coefficient = 0.75).

This is because steroids are more similar in structure compared to amino acids.

Thus, by analysing the mutual correlations of information characteristics, it is possible to estimate the degree of similarity of molecules of one sample compared to molecules of another sample. These data can be useful in forming training samples for QSAR/QSPR problems.

3.7. Analysis of non-equivalence of vertices of molecular graphs

In this part of the paper, an attempt is made to use the IP concept within the framework of the topological model of a molecule (molecular graph) to analyse the equivalence/non-equivalence of atoms and molecular fragments of structurally homogeneous molecules, in particular, fullerenes. The solution of such a problem may be in demand, for example, to determine the canonical numbering in the molecular graph, to interpret various spectra, to solve various nomenclature problems, etc.

At present, there are quite a few works known that allow to construct lists of equivalent/non-equivalent vertices of graphs [85-94]. The following reasons for the wide interest of chemists in the problem of recognising equivalence classes of vertices in

graphs can be distinguished:
(A) construction of all (or a certain part of) substitutions from the group of automorphisms of a given graph;
(B) recognising isomorphism (or non-isomorphism) for two given graphs;
(C) determination of the canonical numbering of vertices of a given graph (this numbering corresponds to a matrix, the form of which does not depend on the initial numbering of vertices, which allows us to uniquely encode the graph structure on its basis);
(E) finding equivalence classes (orbits of the vertex group of a graph).
All problems A - C are important for chemists (A is necessary for designing derivatives of a given structure; without B and, especially, C, it is impossible to compile and use databases) and all of them are strictly solved by back-track methods, the labour intensity of which is exponential in the general case. The ability to solve the problem D for arbitrary graphs with simultaneous establishment of linear order on the set of equivalence classes (in other words, with unique numbering of found orbits) allows to solve strictly the problem C in polynomial time. On the other hand, the computation of any vertex invariants makes it possible to partition the set of graph vertices into classes.
Most of the proposed algorithms are heuristic, the exact solution for arbitrary graphs is provided only by several later procedures, which include more or less strong heuristics at the initial stage and then combinatorial brute force at the last stage. For strongly regular graphs, all problems are particularly difficult to solve and, to our knowledge, no one has proved the correctness of heuristic procedures for them. The heuristic method proposed in this paper[5] is non-iterative, simple enough and does not require large amounts of machine time.
The developed FE concept is directly applicable to the recognition of geometrical equivalence/non-equivalence of atoms and molecular fragments. For the corresponding evaluations within the topological model a modification of the FE concept is necessary. In fact, it is a matter of applying
of the above concept to topological models of compounds - molecular graphs. For molecular graphs, the only possible "channel" of information transfer is a system of edges. Thus, in a topological molecular model, information from one atom to another can be transferred only by chemical bonds (edges of the graph), such as, for example:

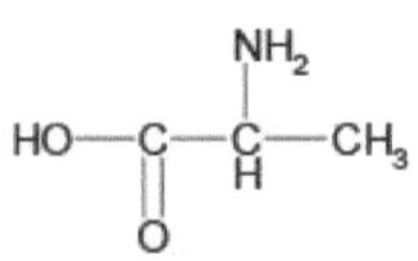

To estimate information potentials (IP) on atoms within the topological model, a modified calculation scheme corresponding to 2D space was used [58]. It should be noted that there are two variants of calculation of topological atomic information potentials:
1) MI method - the shortest distances between atoms are used (the number of paths (m) between each pair of atoms is 1),

[5] In particular, the non-iterative approach proposed in this paper correctly defines three orbits in the skeleton of the kunean, which has been a "stumbling block" for many heuristic methods, including iterative ones.

2) M2 method - all possible paths of information propagation along the graph edges between each pair of atoms (m >1) are used:

$$IP_i = w_i \cdot \sum_{j=1, j\neq i}^{n} \left(\frac{\sum^{m} lb\left(\frac{r}{2R_{ij}+1}\right)}{m} \right), \quad \text{where} \quad w_i = p_i / \sum_{i=1}^{n} p_i, \quad lb = \log_2,$$

where m is the number of paths for each pair of atoms, n is the number of atoms in the molecule, Rij is the number of bonds between the i-th and j-th atoms (path length), w_i *is* a weight parameter describing some property *(p)* of atoms (charge, lipophilicity, polarisability, etc.), (w_i =1 in the case of an unweighted information field), r is a tuning parameter of the model (in the present work we used the value r = 256).

To calculate the information potential within the M2 method, a path in a molecular graph is defined as a non-return route ("self-avoiding walk"), which is a sequence of adjacent vertices, with each vertex not meeting more than once. Fig. 3.21 (b) shows an example of a tree of all possible paths for vertex 1 of naphthalene. By analysing this tree, we can easily compute the lengths of all possible paths from vertex 1 to the rest of the vertices in the graph and thus compute the value of IP1 (Table 1). By repeating the same procedure for all naphthalene atoms, we can obtain the IP values at each atom (Fig. 1c).

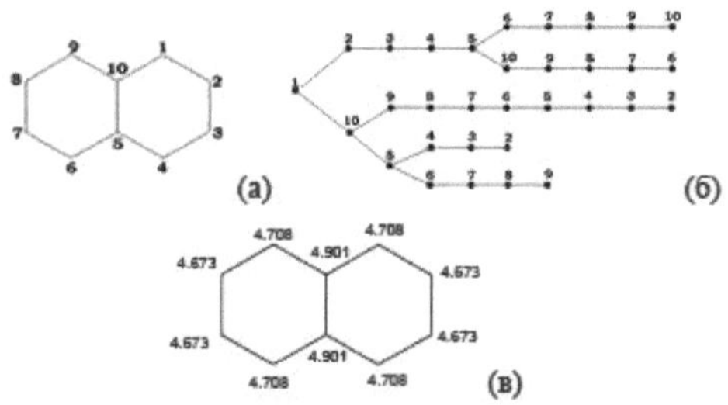

Figure 3.21. Example of IP calculation, for a naphthalene molecule (M2 method).

The lengths of all paths from vertex 1 to the other vertices in mothballs.

Vertex, i	Lengths of all paths to vertex 1[a]	IP1i
2	(1 ,9 ,5)	4.903
3	(2 ,8 ,4)	4.807
4	(3 ,7 ,3)	4.826
5	(4 ,6 ,2)	4.936
6	(5 ,9 ,5 ,3)	4.506
7	(6 ,8 ,4 ,4)	4.468
8	(7 ,7 ,3 ,5)	4.480
9	(8 ,6 ,2 ,6)	4.547
10	(9 ,5 ,1)	4.903
IP1 =.		**4.708**

[a] The lengths of all paths from the i-th vertex to vertex 1 are given in brackets. The number of digits in brackets is the number of paths (the value m in the formula for determining the information potential).

It is important to note that the number of all possible paths between atoms increases exponentially with the number of cycles in the molecule, which leads to significant computational difficulties for structures containing a large number of structurally homogeneous fragments. The analysis of such structures is possible if not all possible paths between atoms are considered, but only those whose lengths are less than the critical length (LCRJT). The values of the path lengths above the molecular graph diameter (D)[6] by n are used as LCRIT: $L_{CRIT} = D + n,\ n \in N$. As a result of using this assumption, the computational speed increases significantly (complexity is less than exponential) .[7]

In this paper, another variant of the topological model of the information field (M3) was considered, in which the attenuation coefficient (k)[8] is used to account for the loss of information during transmission by communication. The degree of the vertex d of the graph is used to account for the branching of the "information flow". The total amount of information that is transmitted per atom is determined by taking into account the informational influence of all other atoms on it. For each atom, the total amount of information is calculated using the following formula:

$IP(i) = I_0(i) + \sum_{j \neq i}^{n} w_j \cdot \left[k \cdot d_j^{-1}(10 - lb(r_{ij}+1))\right]$, where $w_i = p_i / \sum_{i=1}^{n} p_i$, $I_0(i) = lb(w/w_i)$, $I_0(i)$ – is the IP value on atom i without taking into account the influence of the environment; n-total number of atoms in the structure; rij - topological distance from the current i-ro atom to the j-ro atom.

The previously developed information field model (M4), which takes into account the geometry of molecules, was used for comparison with topological models (M1-M3) [58].

To analyse the efficiency of the above approaches, a test problem was solved. Triphenylene derivatives were used as a model example (Fig. 3.22). Previously, the non-equivalence of the A and B vertices of molecular graphs was analysed using the method of graph potentials [95]. The authors showed that topologically non-equivalent vertices A and B are isospectral, and such methods of graph vertex classification as Morgan's algorithm for computing extended connectivity, eigenvector and characteristic polynomial methods do not distinguish them. The method of potentials of the second kind proposed in [95] allows us to distinguish vertices A and B. However, given the use of numerous matrix operations in the method of graph potentials, we can assume that its use for large structurally homogeneous systems such as fullerenes, nanotubes, etc., may cause significant computational difficulties.

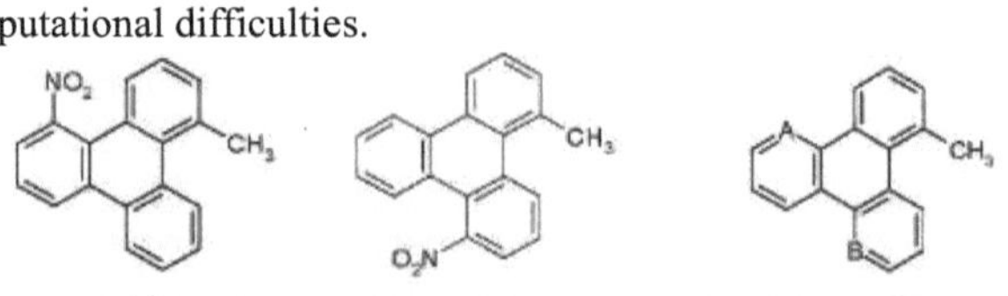

Потенциалы II рода	M1	M2	M3
$U_A = 2.46966$	$IP_A = 5.3336775$	$IP_A = 4.2687811$	$IP_A = 2.23912$
$U_B = 2.46957$	$IP_B = 5.3336779$	$IP_B = 4.2687814$	$IP_B = 2.23812$

Fig.3.22. Results of analysis of non-equivalence of vertices (A and B) of molecular graphs

[6] The diameter of a graph is the maximum distance between vertices for all pairs of vertices.

[7] For example, for the fullerene C32, the calculation time with full path search was 25 minutes, while using the constraint, the calculation took only a second

[8] k is the tuning parameter of the model, k = 0.99

of triphenylene derivatives.
The implementation of the above information field algorithms is much simpler and, therefore, can be applied to the analysis of large structurally homogeneous molecular systems. The information field algorithms also cope with the above test problem. As can be seen from the table (Fig. 3.22), the IPs evaluated by M1-M3 for the non-equivalent vertices A and B are different. Note that the M3 method is the most sensitive in this case, which resulted in the maximum difference between the IPs of vertices A and B. M2 compared to M1 did not show higher sensitivity when analysing the non-equivalence of the above vertices, probably because the number of alternative paths between different pairs of vertices of the considered molecular graphs is relatively small.
Further, an attempt was made to investigate the possibilities of the developed approaches to recognise the non-equivalence of atoms of more complex structurally homogeneous molecular systems. Fullerenes C32, C76, C78 [96], the structural analysis of which was carried out earlier [97] by one of the co-authors of this work, were considered as objects of study. A classification system of fullerenes was developed based on differentiation of vertices in terms of their occurrence in cycles of different sizes [97]. Vertices of the same type can form several non-overlapping clusters. The maximum number of equivalent atoms (EA) is determined by the symmetry group. These values can be compared with the maximum number of equivalent atoms calculated by M1-M4 (Table 3.3).
As can be seen from Table 3.3, the classification system based on the type of vertex articulation does not allow us to fully distinguish non-equivalent atoms for the studied fullerenes. The use of topological methods taking into account the minimum information propagation path (M1 and M3) allows us to distinguish already most of the non-equivalent atoms. The maximum differentiation of atoms is obtained by using M2 and M4, and at the same time the maximum number of equivalent atoms, as one would expect, corresponds to the EA value.
In addition, M2 and M4 separate the atoms into groups in the same way (Figs. 3.23a, 3.24, 3.26). This may indicate that the geometrical inequivalence of atoms is determined by their topological inequivalence.

Estimation of the maximum number of equivalent atoms (EA) for the investigated fullerenes

molecule	Symmetry group	*EA* in a general position	*EA* according to the vertex classification system [10]	*EA* estimated by the method			
				M1	M2	M3	M4
C32	D3	6	24	12	6	12	6
C76	D2	4	60	8	4	8	4
C78	D3	6	60	6	6	6	6

The above figures also show that M2 and M4 allow to remove the "degeneracy" in the largest groups of atoms formed as a result of M1 and M3: cluster C for fullerene C32 (Fig. 3.23), clusters O and P for C76 (Fig. 3.25). For fullerene C78, all considered approaches (M1-M4) give the same pattern of atom partitioning into equivalence classes (Fig. 3.26).
Thus, the obtained results indicate that the topological model of the IP can be used to analyse the non-equivalence of atoms or groups of various molecular systems, including

such structurally homogeneous compounds as fullerenes. As one might expect, the most effective implementations of the topological model of the FE are those that take into account not only paths of minimum length between pairs of vertices, but also other paths of information propagation along the edges of the corresponding molecular graphs, while this heuristic approach is quite simple, is not iterative and does not require large expenditures of machine time.

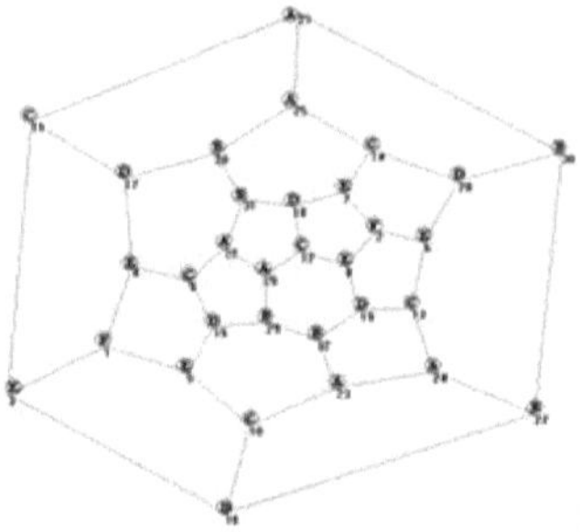

a)

	group	atom numbers	IP (M2)	IP (M4)
1	A	21, 22, 23, 24, 25, 26	4.8846	16.5502
2	B	27, 28, 29, 30, 31, 32	4.8838	16.5502
3	C	9, 10, 11, 12, 13, 14	4.8781	16.5050
4	D	15, 16, 17, 18, 19, 20	4.8732	16.5245
5	E	3, 4, 5, 6, 7, 8	4.8731	16.4604
6	F	1, 2	4.7205	16.4108

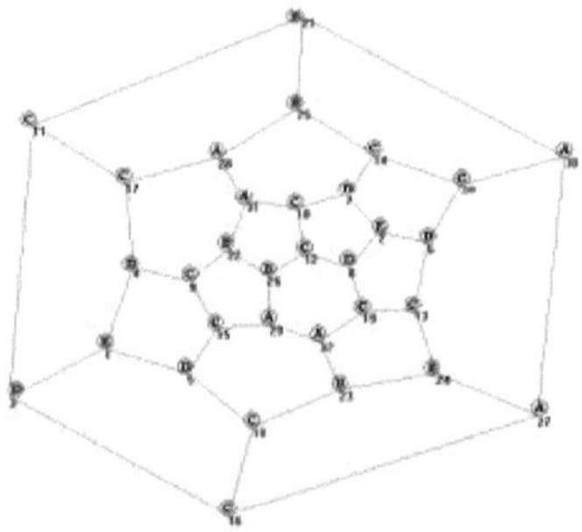

б)

	group	atom numbers	IP (M1)	IP (M3)
1	A	27, 28, 29, 30, 31, 32	5.4354	87.638
2	B	21, 22, 23, 24, 25, 26	5.4144	87.350
3	C	9, 10, 11, 12, 13, 14, 15, 16, 17, 18, 19, 20	5.3973	87.116
4	D	3, 4, 5, 6, 7, 8	5.3895	87.009
5	E	1, 2	5.3447	86.394

Figure 3.23: Schlegel diagram, atom groups and IP values obtained under: a) M2 and M4, b) M1 and M3, for C32.

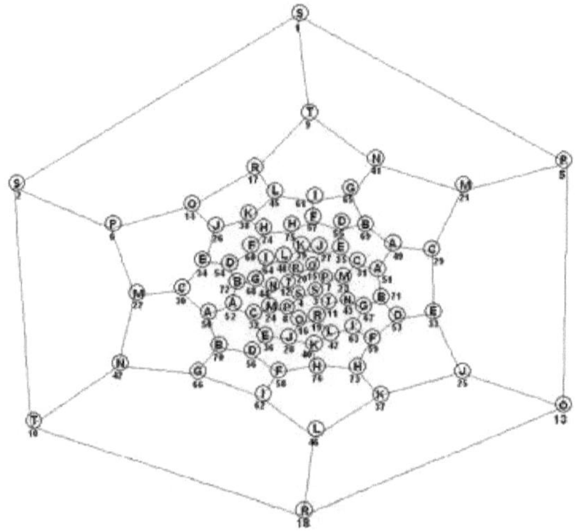

group	atom numbers	IP (M2)	IP (M4)
A	49, 50, 51, 52	4.0565	14.8717
B	69, 70, 71, 72	4.0550	14.7740
C	29, 30, 31, 32	4.0530	14.8396
D	53, 54, 55, 56	4.0051	14.7711
E	33, 34, 35, 36	4.0033	14.8094
F	57, 58, 59, 60	4.0016	14.7268
G	65, 66, 67, 68	4.0006	14.6958
H	73, 74, 75, 76	4.0003	14.7606
I	61, 62, 63, 64	3.9997	14.6756
J	25, 26, 27, 28	3.9990	14.7182
K	37, 38, 39, 40	3.9985	14.7099
L	45, 46, 47, 48	3.9971	14.6561
M	21, 22, 23, 24	3.9473	14.7160
N	41, 42, 43, 44	3.9468	14.6688
O	13, 14, 15, 16	3.9450	14.6539
P	5, 6, 7, 8	3.9443	14.6519
R	17, 18, 19, 20	3.9425	14.6262
S	1, 2, 3, 4	3.9400	14.6093
T	9, 10, 11, 12	3.8914	14.6187

Fig. 3.24. Schlegel diagram, atom groups and IP values obtained in the framework of M2 and M4, for C76.

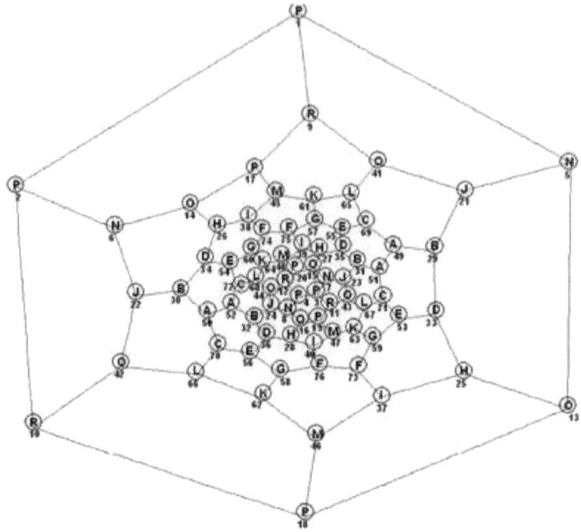

group	atom numbers	IP (M1)	IP(M3)
A	49, 50, 51, 52	4.7312	188.725
B	29, 30, 31, 32	4.7192	188.316
C	69, 70, 71, 72	4.7120	188.083
D	33, 34, 35, 36	4.7104	188.022
E	53, 54, 55, 56	4.7079	187.943
F	73, 74, 75, 76	4.6999	187.667
G	57, 58, 59, 60	4.6974	187.589
H	25, 26, 27, 28	4.6929	187.434
I	37, 38, 39, 40	4.6923	187.415
J	21, 22, 23, 24	4.6903	187.348
K	61, 62, 63, 64	4.6852	187.180
L	65, 66, 67, 68	4.6806	187.025
M	45, 46, 47, 48	4.6800	187.006
N	5, 6, 7, 8	4.6736	186.785
O	13, 14, 15, 16, 41, 42, 43, 44	4.6714	186.712
P	1, 2, 3, 4, 17, 18, 19, 20	4.6609	186.358
R	9, 10, 11, 12	4.6590	186.357

Fig. 3.25: Schlegel diagram, atom groups and IP values obtained within the M1 and M3, for C76.

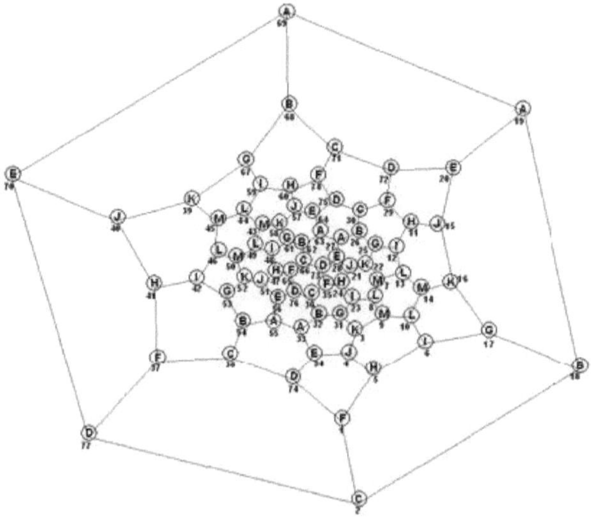

group	atom numbers	IP (M1)	IP (M2)	IP (M3)	IP (M4)
A	19, 27, 33, 55, 63, 69	4.7039	4.0004	192.818	14.8028
B	18, 26, 32, 54, 62, 68	4.6943	3.9978	192.483	14.7758
C	2, 30, 36, 38, 66, 71	4.6920	4.0000	192.409	14.7529
D	72, 73, 74, 75, 76, 77	4.6896	4.0013	192.331	14.7138
E	20, 28, 34, 56, 64, 70	4.6852	3.9993	192.176	14.7076
F	1, 29, 35, 37, 65, 78	4.6800	3.9989	191.996	14.6834
G	17, 25, 31, 53, 61, 67	4.6660	3.9431	191.508	14.6561
H	5, 11, 24, 41, 47, 60	4.6591	3.9457	191.273	14.6212
I	6, 12, 23, 42, 48, 59	4.6561	3.9418	191.173	14.5998
J	4, 15, 21, 40, 51, 57	4.6547	3.9458	191.118	14.6311
K	3, 16, 22, 39, 52, 58	4.6477	3.8927	190.878	14.6035
L	8, 10, 13, 44, 46, 49	4.6375	3.88673	190.524	14.5516
M	7, 9, 14, 43, 45, 50	4.6354	3.88669	190.450	14.5527

Figure 3.26: Schlegel diagram, atom groups and IP values obtained under M1-M4 for C78.

CHAPTER 4

APPLICATIONS OF THE MOLECULAR INFORMATION FIELD CONCEPT FOR SOLVING STRUCTURE-PROPERTY PROBLEMS

Quantitative Structure-Property Relationships for Organic Compounds (QSAR/QSPR) studies are now receiving quite a lot of attention in order to predict the behaviour of new molecules using data obtained from the analysis of the properties of already tested molecules.

The increasing interest in this problem is primarily due to the fact that the identification of such relationships contributes to a better understanding of the mechanisms of functioning of new materials, reagents, and drugs. The use of computer technologies opens up the possibility to carry out not only preliminary screening (screening) but also molecular design of organic substances exhibiting one or another property on the basis of the revealed regularities. All this makes it possible to significantly reduce the enormous expenditure of time and resources required for the development of new substances with a set of specified properties.

There are quite a large number of methods and models for solving "structure-property" problems. The simplest models are based on the use of substituent constants, topological indices of molecules (EMMA [98], CODES S A [99]), and information on the number of structural fragments of one or another type (HQSAR method [100, 101]). In more complex 3D QSAR methods (e.g., CoMFA [102, 103]), spatial structure characteristics play the main role.

Using the concept of information field developed by us, a number of problems on modelling biological activity and some other properties of molecules were solved on the basis of IP parameters weighted by various properties of molecular atoms. The search for quantitative structure-property relationships was carried out using the multiple linear regression analysis (MLRA) method [104, 105].

4.1. Study of steroid affinity for corticosteroid- and testosterone-binding globulin

To analyse the effectiveness of the developed approach, a series of steroids were studied (Appendix B). The data on binding of these compounds to corticosteroid- and testosterone-binding globulins (CRCG and TSG) were considered. The studied molecules belong to hormones that stimulate the development and functioning of the genital system, the development of secondary sexual characteristics, regulate growth and some aspects of human and animal behaviour. Acting as androgens, steroids flow through the cellular environment into the cytoplasm, where they bind to the receptor into an active complex that is transported to the cell nucleus.

This series of compounds is well studied and is often used to test the performance of developed QSAR models [102, 106, 107]. In particular, quite satisfactory results were obtained by the CoMFA method: the square of the correlation coefficient (R^2) between the calculated and predicted binding values was 0.897 and 0.873 for CRC and TSG, respectively [102].

Previously, the structure-affinity problem of steroids for CRC and TSH for a number of compounds was solved in [108] using harmonic, local, integral, and field structural parameters. Based on the QSAR models obtained, it was concluded that an important factor affecting the biological activity of steroids is the distribution of oxygen atoms in the molecular structure. Hence, one of the hypotheses arose that the formation of intermolecular hydrogen bonding significantly affects the interaction of steroid molecules with CRCs and TSGs.
In the present work, we have attempted to solve this QSAR problem using only the FE parameters.
The studied steroids were considered within the framework of the molecular lattice model [108, 109]. The structures of the compounds (21 molecules) were optimised using the molecular mechanics method (MM+ force field) [110] and then aligned according to carbon atoms 5-14 underlying the rigid framework of steroid molecules:

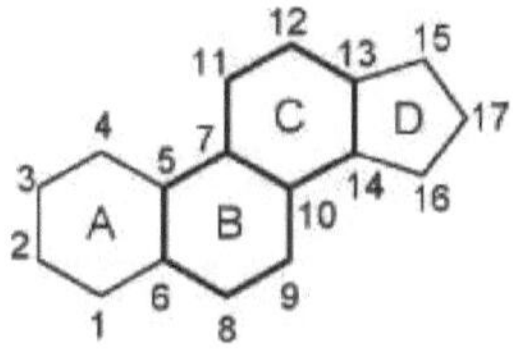

After that, the investigated structures were placed in a cubic lattice with a cell size of 2.0 A. The analysed region of space was a cubic lattice consisting of 1000 cells (10x10x10), in the centre of which the steroid molecules were placed. In each cell of this region, the IP parameters were evaluated: potential, gradient, and gradient field divergence, weighted by stereochemical "charges", lipophilicity, refraction, masses, polarisability, and electronegativity. In total, about 20 thousand structural parameters were considered. After screening out mutually correlated structural parameters by stepwise regression, we were able to obtain three-parameter equations linking the parameters of IP steroids to their affinity for CRC and TSH (Sections 4.1.2. and 4.1.3.).
In addition to scalar FE parameters, vector FE parameters of potential hydrogen bonding potentials were calculated (Section 4.1.1), which were also used in the construction of QSAR models.

4.1.1. Construction of vector IPs of potential hydrogen bonds.

As it was already mentioned in Section 2.3.2. the vector information field model can be constructed if the object under study is represented as a system of vectors. It seems to us that the vector information field model is a convenient tool for modelling hydrogen bonds [111, 112].
Hydrogen bond is a special type of three-centre chemical bond of the X-II...Y type, in which the central H atom, connected by a covalent bond with an electronegative atom X (C, N, O, S), forms an additional bond with an atom Y (N, O, S) having an unshared electron pair directed along the line of this bond. The hydrogen atom plays the role of a bridge connecting the fragments X and Y. That is, for the formation of a hydrogen bond, two partners are required, one of which acts as an acceptor of the hydrogen bond (fragment Y) and the other as a donor (fragment X-H). In hydrogen bonding, the linear

configuration of atoms is energetically advantageous, so it is possible to model by vectors the proton displacement to the electron donor along the bonding line:

H-bond donor *H-bond acceptor* *H-bond acceptor*

Any molecule that has fragments capable of forming hydrogen bonds can be represented by a system of vectors reflecting their ability to be hydrogen bond donors and/or acceptors. The lengths of such vectors can be characterised by the values of acid (electron acceptor) P_i and basic (electron donor) E_i factors proposed by A.V. Johansen in [113]. The calculation of these factors follows from the idea of the constancy of acid and basic functions of molecules, the mutual independence of these functions, which predetermine the strength of H-bonds. Each molecule in associates with other molecules makes its constant contribution. Quantitatively this representation is expressed by the rule of product of parameters (factors) of "acids" and "bases" in hydrogen bonds - the "rule of factors":

$$\Delta H_{ij} = \Delta H_{11} P_i E_j,$$

where the proportionality coefficient *Sh11*- enthalpy of an arbitrary standard H-complex whose partners have P1 = *£) = 1*.

Phenol and diethyl ether were proposed as a standard system, for which ΔH_{11} = 5.3 kcal/mol. The values of acid (electron acceptor) *Pi* and basic (electron donor) *Ei* - Johansen factors, which were used in this work, are given in Table 4.1.

Table 4.1.

Values of acid (electron acceptor) *Pi* and basic (electron donor) *Ei* - factors of some fragments of molecules according to Johansen's methods

	molecular fragment	factor value (vector length)
H-bond donor	**O-H→**	*Pi* =0.63
H-bond acceptor	**>C=O←**	*Ei* =0.93
H-bond acceptor	**>O←**	*Ei* =1.00
H-bond acceptor	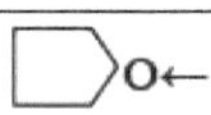	*Ei* =1.04

Thus, representing a molecule as a system of vectors of a certain length reflecting donor and acceptor fragments of the molecule, it is possible to construct a vector IP characterising the possibility of formation of potential hydrogen bonds. We are aware that intermolecular hydrogen bonding is a rather short-range interaction and is realised at distances smaller than the sum of van der Waals radii of H and Y atoms. Nevertheless, modelling the hydrogen bonding IP for molecules at large distances, we take into account the fact that interacting molecules seem to "feel" each other, and even before the formation of a real hydrogen bond "prepare" for the attractive interaction, orienting themselves in space in an appropriate way.

4.1.2. Corticosteroid - binding globulin.

The search for dependencies describing the structure-affinity relationship of steroids to

CRC was carried out separately for the group including such scalar IP parameters as divergence, gradient, and potential IP (Model I) and for the group consisting only of vector IP parameters of potential hydrogen bonds (Model II). A QSAR model including the structural parameters of both groups was also constructed (Model III). As a result, we were able to obtain three roughly equivalent models I, II, III for these cases. Table 4.2 presents for these models the parameters of the correlation equations describing the relationship between the characteristics of the information fields of the studied steroid molecules and their affinity for corticosteroid-binding globulin, and Table 4.3 presents the observed and predicted values of biological activity according to the three obtained models.

In the first case (Model I), when all scalar FE parameters were used, about 18 thousand structural parameters were considered. After screening out mutually correlated structural parameters (at $r > 0.9$), the number of parameters was reduced to 216. A three-parameter equation (Table 4.2.) was obtained by the stepwise regression method, which involved the divergence parameters of the gradient field of lipophilicity- and electronegativity-weighted IPs and the divergence parameter of the gradient field of unweighted IPs.

Fig. 4.1 (a) shows the cubic cells of the information parameters used in Model 1. Positive influence on the affinity of steroid molecules to CRC will be exerted by electronegative groups in the region of D ring (see Scheme 4.1.) (cell 49), as well as the presence of substituents in the region of D and B rings (cell 12). Negative influence on the interaction of steroids with CRC is exerted by the presence of lipophilic fragments in the region of ring A atom 2 (cell 354).

For the construction of Model II, only the parameters of vector IPs of potential hydrogen bonds were used. In Sect. 4.1.1. (Table 4.1) presents the values of donor and acceptor Johansen factors used in this model for the corresponding fragments.

Characteristics of QSAR models describing the relationship between the structure of steroid molecules and their affinity for corticosteroid-binding globulin

A = ax + by + cz + d

Characteristics of the model	Model I	Model II	Model III
Number of structural parameters before screening / after screening	18000/216	2000/193	20000/349
a	-0.55	-2.6	-0.61
x	DL(354)	DHyd(526)	DL(19)
b	0.11	1.8	45.0
y	DEn(49)	DHyd(687)	DGeom(220)
c	2.23	1.9	-1.0
z	DGeom(12)	DHyd(616)	DHyd(661)
d	-12.3	-4.0	16.0
R^2	0.92	0.88	0.93
CVR^2	0.89	0.81	0.91
S	0.36	0.45	0.33
F	65.7	39.6	78.0

D_L (354) and D_L (19) are the divergence of the gradient field of lipophilicity-weighted information fields (in cells 354 and 19);
D_{Hyd} (526), D_{Hyd} (687), D_{Hyd} (616), D_{Hyd} (661) is the divergence of VIPs characterising potential hydrogen bonds in cells 526, 687, 616 and 661, respectively;
D_{En} (49) is the divergence of the gradient field of the information fields weighted by electronegativity (in cell 49);
D_{Geom} (12) and D_{Geom} (220) are the divergence of the unweighted FE gradient field (in cells 12 and 220);
R - correlation coefficient;
CVR is the correlation coefficient, under rolling control conditions;
S is the standard error of the estimate;
F is Fisher's criterion.

Observed and calculated values of the affinity of steroid molecules for the corticosteroid-binding globulin (CBG).

molecule number	observed affinity values (A)	Predicted affinity values (A)		
		Model I	Model II	Model III
1	-6.28	-6.39	-6.16	-6.60
2	-7.65	-7.15	-7.25	-7.60
3	-7.88	-8.15	-7.61	-7.96
4	-5.92	-6.34	-6.48	-6.28
5	-5.00	-4.72	-5.23	-5.00
6	-5.00	-5.15	-5.25	-5.29
7	-5.00	-5.22	-4.50	-5.00
8	-5.26	-5.05	-4.62	-5.28
9	-5.26	-5.49	-5.48	-5.47
10	-5.00	-5.45	-5.51	-5.26
11	-7.38	-6.85	-7.13	-7.17
12	-5.00	-4.73	-5.32	-5.44
13	-7.74	-7.45	-7.50	-7.32
14	-6.72	-6.39	-6.72	-6.40
15	-5.00	-4.65	-4.95	-4.46
16	-5.76	-6.43	-6.22	-6.10
17	-5.61	-5.57	-5.98	-5.45
18	-7.88	-7.74	-7.72	-7.81
19	-7.88	-7.92	-7.14	-7.55
20	-6.89	-7.05	-7.65	-7.16
21	-5.00	-5.24	-4.70	-4.52

A total of 2 thousand structural parameters were considered. After accounting for reciprocal

correlations ($r > 0.9$), the number of parameters decreased to 193. QSAR analysis resulted in a three-parameter regression equation (Table 4.2.).

The obtained results indicate (Fig. 4.1.(6)) that the presence of hydrogen bond acceptor groups in the D ring (cell 687) and in the A ring (cell 616) will favour the manifestation of the studied biological activity, while the presence of hydroxyl groups (potential H-bond donors) in the region of the A ring (cell 526) will, on the contrary, hinder the binding of steroids to CRC. In this and the following figures, the vectors in the cells show the possible location of the target fragments that can form hydrogen bonds with the fragments of of steroid molecules. The white colour of the vector denotes a hydrogen atom, and black denotes an electronegative atom linked by a covalent bond to the hydrogen atom of the target.

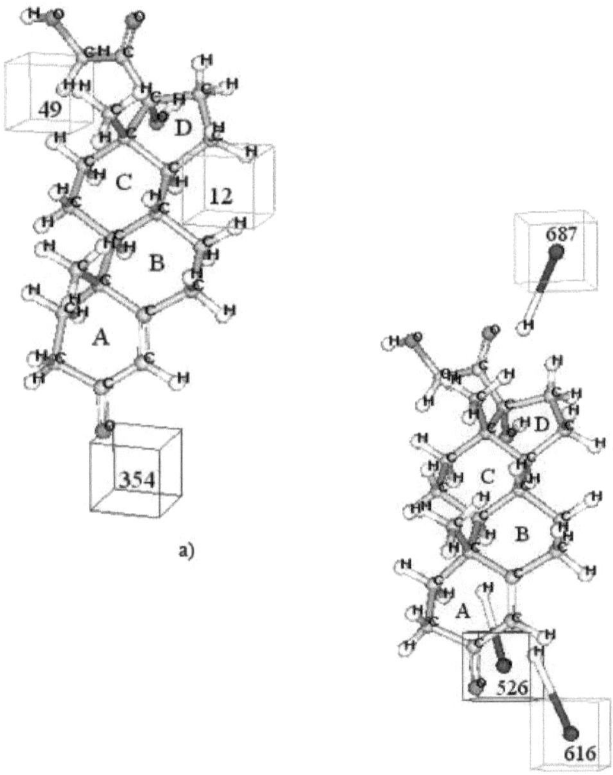

Fig.4.1 Cubic cells in which the parameters of the information fields of steroids determine their affinity for CRC:

a) Model I; b) Model II.

A three-parameter regression equation was also obtained for model III (Table 4.2.), which involved the divergence parameters of the gradient field of the IP, lipophilicity-weighted and unweighted IP. It should be noted that in addition to the parameters of scalar IPs, this model also uses the divergence parameter of vector IPs of potential hydrogen bonds.

As can be seen from Fig. 4.2, the presence of lipophilic fragments in the region of ring B (cell 19) will have a negative effect on the interaction of steroids with CRC. The presence of substituents in the region of rings B and D (cell 220) will on the contrary favour the activity.

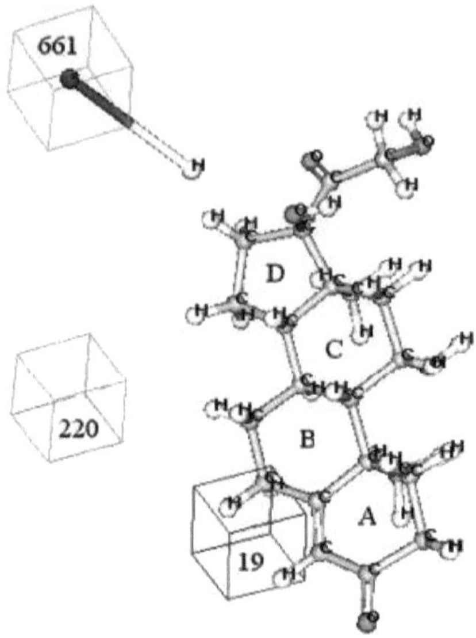

Fig.4.2 Cubic cells in which the parameters of the information fields of steroids determine their affinity for CRC (Model III).

As in the previous case (Model II), the presence of potential hydrogen bond acceptor groups in the D ring region (cell 661) will also contribute to the biological activity under study.

Thus, using the concept of FE molecules, it was possible to obtain quite adequate QSAR models for steroid binding to CRC. It can be noted that model III is statistically superior (R^2 =0.93, CVR^2 =0.91) to the model obtained by CoMFA method (R^2 =0.90, CVR^2 =0.66) [102]. In addition, this model supports the hypothesis of the important role of intermolecular hydrogen bonds for the binding of steroid molecules to CRCs.

4.1.3. Testosterone-binding globulin.

As in the case of the study of steroid affinity to CRC, the dependencies describing the structure-steroid affinity relationship to testosterone-binding globulin (TSG) were searched separately for the group including only scalar FE parameters (Model IV) and for the group consisting only of vector FE parameters of potential hydrogen bonds (Model V). A QSAR model including structural parameters of scalar IPs and vector IPs of potential hydrogen bonds was also constructed (Model VI). This resulted in three roughly equivalent models IV, V, VI. Table 4.4. for these models presents the parameters of the correlation equations describing the relationship between the characteristics of the information fields and the affinity of steroids for TSH, and Table 4.5. presents the observed and predicted values of biological activity for steroid molecules according to the three obtained QSAR models.

A three-parameter regression equation was obtained for model IV (Table 4.4), which involved the divergence parameters of the gradient field of lipophilicity- and atomic mass-weighted IPs and the divergence parameter of the gradient field of unweighted IPs.

Characteristics of QSAR models describing the relationship between the structure of

steroid molecules and their affinity for testosterone-binding globulin

A = ax + by + cz + d

Model characteristics	Model IV	Model V	Model VI
Number of structural parameters before screening / after screening	18000/234	2000/136	20000/342
a	-2.56	1.40	1.16
x	DAW(114)	DHyd(465)	DHyd(365)
b	0.24	-0.32	-1.0
y	DGeom(132)	DHyd(584)	DGeom(61)
c	0.05	-0.50	0.73
z	DL(25)	DHyd(524)	DL(144)
d	7.83	-8.51	-12.5
R^2	0.94	0.94	0.95
CVR^2	0.91	0.9	0.92
S	0.32	0.31	0.29
F	87.9	97.3	112.1

D_{AW} (114) is the divergence of the gradient field of the information fields weighted by atomic masses (in cell 114);

D_{Hyd} (465), D_{Hyd} (584), D_{Hyd} (524), D_{Hyd} (365) is the divergence of vector FEs characterising potential hydrogen bonds in cells 465, 584, 524 and 365, respectively;

D_{Geom} (132), D_{Geom} (61) is the divergence of the gradient field of the unweighted information fields (in cells 132 and 61);

D_L (144), D_L (25) is the divergence of the gradient field of lipophilicity-weighted information fields (in cells 144 and 25);

Observed and calculated affinity values of steroid molecules for testosterone-binding globulin (TSG).

molecule number	observed affinity values (A)	Predicted affinity values (A)		
		Model IV	Model V	Model VI
1	-5.32	-5.31	-5.70	-5.41
2	-7.38	-7.19	-7.18	-7.28
3	-7.2	-7.17	-7.23	-6.98
4	-9.74	-9.86	-9.31	-9.11
5	-8.83	-9.15	-9.05	-8.57
6	-6.63	-6.45	-6.77	-6.95
7	-8.18	-8.10	-7.64	-8.07
8	-6.15	-7.04	-6.33	-6.12
9	-7.15	-6.89	-7.10	-7.18
10	-6.36	-6.55	-6.04	-6.08
11	-6.94	-7.34	-7.25	-7.21
12	-9.11	-9.10	-9.45	-9.34
13	-7	-7.02	-7.17	-7.28
14	-9.2	-8.79	-9.29	-9.22
15	-9.18	-9.06	-9.13	-9.19
16	-7.46	-7.18	-7.49	-7.88

17	-7.15	-7.30	-7.40	-7.59
18	-6.34	-6.25	-6.67	-6.20
19	-6.2	-6.23	-6.01	-5.93
20	-6.43	-6.48	-6.23	-6.41
21	-7.82	-7.33	-7.33	-7.76

Fig. 4.3.(a) shows the cubic cells of the information parameters used in Model IV. As can be seen from the figure, the presence of substituents in the D ring region (cell 132) and lipophilic fragments (cell 25) will have a positive effect on the interaction of steroids with TSH. The presence of substituents in the region of ring C (cell 114) will on the contrary negatively affect the affinity of steroid molecules to TSH.

To build Model V, only the parameters of vector IPs of potential hydrogen bonds were used. In this QSAR model we used the values of donor and acceptor Johansen factors presented in Table 4.1. for fragments of molecules capable of acting as donors and/or acceptors of hydrogen bonds.

The results obtained indicate (Fig. 4.3.(6)) that the presence of hydroxyl groups in ring D (cell 465), which are hydrogen bond donors, will favour the biological activity under study, while the presence of groups acting as potential hydrogen bond acceptors in the region of ring B (cell 524) and in the region of ring D (cell 584) will conversely inhibit the binding of steroids to TSH.

Comparing the results of analyses of models II and V, one can notice a difference in the interaction of steroids with CRCs and TSGs. In the region of the D ring (cells 687 and 584) (Fig.4.1.(b) and Fig.4.3.(b)), the presence of hydrogen bond acceptor groups will positively influence the binding of steroids to corticosteroid-binding globulin, while the presence of hydrogen bond donor groups in the region of this ring will favour the affinity of steroids to testosterone-binding globulin.

A three-parameter regression equation was derived for the VI model (Table 4.4.), involving the divergence parameters of the lipophilicity-weighted IP gradient field, unweighted IP, unweighted IP, and the divergence parameter of vector IP of potential hydrogen bonds.

As can be seen from Fig. 4.4, the presence of lipophilic fragments in the region of rings A and B (cell 144), as well as the presence of hydroxyl groups (potential hydrogen bond donors) in the region of ring D (cell 365) will favour the binding of steroids to TSH. The presence of substituents in the region of ring C (cell 61) will have a negative effect on the manifestation of the studied biological activity of steroid molecules.

Thus, using the concept of FE molecules, it was possible to obtain quite satisfactory QSAR models for the affinity of steroid molecules for TSH as well.

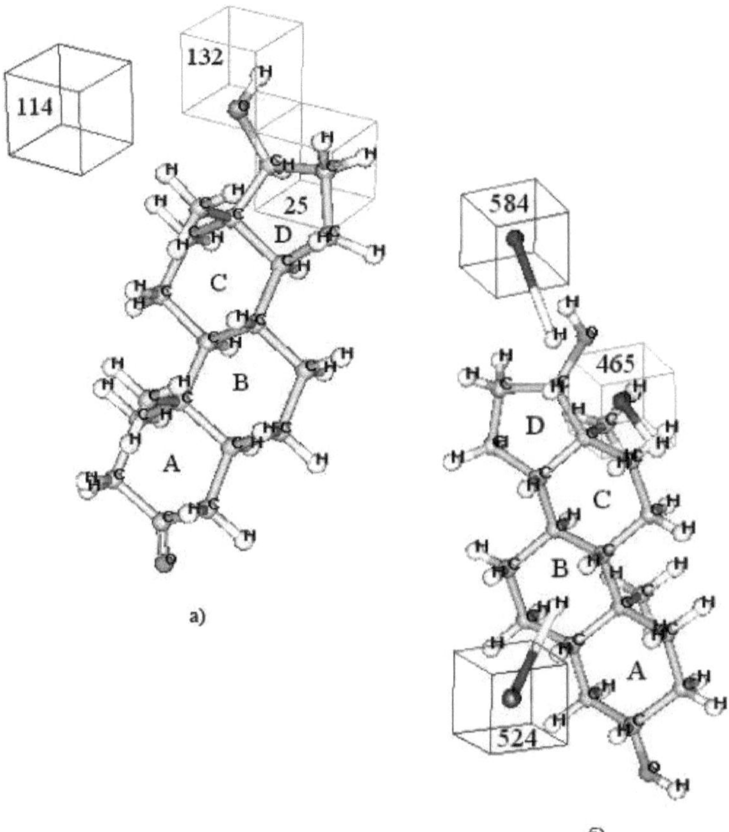

Fig.4.3 Cubic cells in which information parameters of steroids determine their affinity to TSH: a) by Model IV; b) by Model V.

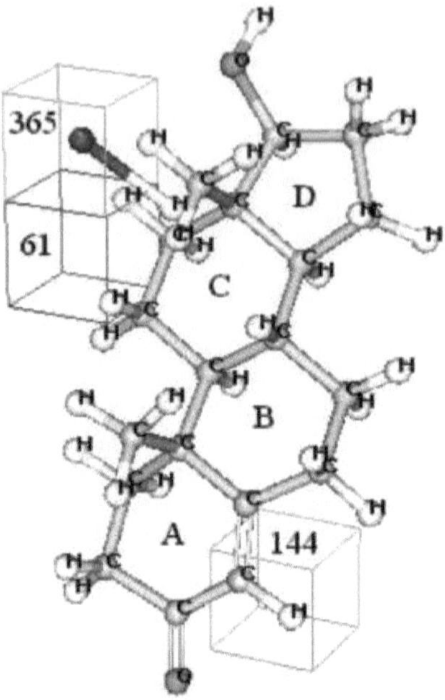

Figure 4.4 Cubic cells in which the parameters of the information fields of steroids determine their affinity for TSH (Model VI).

It can be noted that the VI model is statistically superior to the model obtained by the CoMFA method (R^2 =0.87, CVR^2 =0.60) [102]. As in the previous case (with CRC), this QSAR model also supports the hypothesis of the important role of intermolecular hydrogen bonds in the binding of steroids to TSH. In addition, based on the data obtained, it is possible to speak about the difference in the binding mechanisms of steroids to the corresponding globulins (CRC and TSH).

4.2 Analysing the influence of the structure ofl ,4-benzdiazepin-2-ones on specific binding to the benzdiazepine receptor

Using the concept of IP we attempted to investigate the effect of the structure of a series (57 molecules) of 1,4-benzdiazepin-2-ones, including a number of commercial preparations (Table 4.6.), on the inhibition of[9] benzdiazepine site (BzR) on the LAMKA receptor family [114].

This series of molecules was studied earlier in [115], where different topological indices were used as structural parameters. The analysis of the structure-activity relationship by the MLR method and the PLS method led the authors to unsatisfactory results. Thus, even after removing the "inconvenient" seven structures from the series of 57 molecules, the correlation coefficients of the best regressions were only R=0.721 for MLR and R=0.754 for the PLS method.

In the present work, the same training sample of 57 molecules was considered within the

inhibition parameter ($LogIC_{50}$): IC_{50} in nmol/L

framework of the molecular lattice model [108]. The structures of the studied compounds were optimised using the molecular mechanics method (MM+ force field) [110] and aligned according to the potential pharmacophore groups >N-, >C=O, >C=N-. The investigated molecules were then placed in a cubic lattice with a cell size of 2.0 A. In each lattice cell, the parameters of scalar and vector IPs weighted by electronegativity, lipophilicity, mass, refraction, van der Waals radii of the corresponding atoms of the molecules were calculated. In addition, vector IP parameters of potential hydrogen bonds were calculated, since 1,4-BD molecules contain fragments capable of acting as hydrogen bond donors and/or acceptors. As a result of QSAR analysis, three equivalent regression models were obtained: VII - using only scalar IP parameters, VIII - using scalar IP parameters and vector IP parameters of potential hydrogen bonds, and IX - using all the above parameters and vector IP parameters reflecting the heterogeneity of information distribution within the molecule (Table 4.7.).

Observed and predicted Log(IC50) values in the series of 1,4-benzdiazepines.

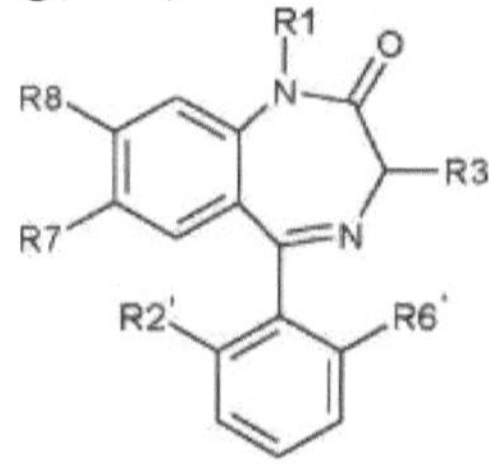

№	R7	R1	R2'	R6'	R3	R8	A3KCn	A7	A8	A9
1	F	H	H	H	H	H	1.602	1.454	1.314	1.698
2	F	Me	H	H	H	H	1.230	1.543	1.342	1.691
3	F	H	F	H	H	H	0.869	0.687	0.548	1.039
4	F	Me	F	H	H	H	0.708	0.978	0.650	1.058
5	Cl	H	H	H	H	H	0.973	1.220	1.223	0.76
6	Cl	Me	H	H	H	H	0.908	1.232	1.314	0.83
7	Cl	H	F	H	H	H	0.301	0.494	0.574	0.322
8	Cl	H	Cl	H	H	H	0.255	0.492	0.062	0.193
9	I	Me	F	H	H	H	0.462	0.114	0.090	0.621
10	CF3	H	H	H	H	H	1.114	1.712	1.550	1.628
11	N3	H	F	H	H	H	0.724	1.192	0.663	0.437
12	NO2	H	H	H	H	H	1.000	1.096	1.386	1.092
13	NO2	H	F	H	H	H	0.176	0.256	0.663	0.52
14	NO2	Me	F	H	H	H	0.580	0.552	0.917	0.636
15	NO2	H	Cl	H	H	H	0.255	0.255	0.063	0.345
16	NO2	Me	Cl	H	H	H	0.342	0.435	0.383	0.392
17	NO2	H	CF3	H	H	H	0.544	0.645	0.458	0.343
18	NHOH	Me	F	H	H	H	1.982	1.651	1.861	2.124
19	NH2	H	H	H	H	H	2.587	2.208	2.774	2.474

20	NH2	Me	H	H	H	H	2.663	2.305	2.852	2.575
21	NH2	Me	F	H	H	H	1.813	1.736	2.089	1.812
22	NH2	H	Cl	H	H	H	1.875	1.350	1.461	1.471
23	CN	Me	H	H	H	H	2.580	1.699	2.082	1.952
24	CN	Me	F	H	H	H	1.477	1.060	1.548	1.433
25	Et	H	H	H	H	H	1.556	1.585	1.468	1.522
26	CH=CH2	H	H	H	H	H	1.380	1.552	1.120	1.439
27	CHO	H	H	H	H	H	1.663	2.213	1.744	1.913
28	COMe	H	F	H	H	H	1.255	1.151	1.086	0.943
29	H	H	H	H	H	H	2.544	1.917	1.970	1.999
30	H	H	F	H	H	H	1.322	1.056	1.302	1.361
№	R7	R1	R2'	R6'	R3	R8	$A^{эксп}$	A7	A8	A9
31	H	Me	F	H	H	H	1.146	1.347	1.313	1.322
32	H	Me	Cl	H	H	H	0.580	1.224	0.813	1.016
33	Cl	CH2CF3	H	H	H	H	1.964	2.246	1.760	2.022
34	NO2	CH2OCH3	H	H	H	H	2.633	2.596	2.551	2.53
35	Cl	$C(CH_3)_3$	H	H	H	H	2.792	2.869	2.567	2.601
36	Cl	(CH2)2OH	F	H	H	H	1.389	1.719	1.208	1.603
37	NO2	$C(CH_3)_3$	F	H	H	H	2.477	2.207	2.263	2.417
38	H	H	F	F	H	H	1.279	0.982	1.441	1.436
39	Cl	H	F	F	H	H	0.204	0.431	0.726	0.314
40	Cl	Me	F	F	H	H	0.613	0.691	0.965	0.312
41	Cl	H	Cl	F	H	H	0.477	0.377	0.270	0.108
42	Cl	H	Cl	Cl	H	H	0.845	0.448	0.662	0.767
43	Cl	Me	Cl	Cl	H	H	0.740	0.585	0.812	0.672
44	Br	Me	F	F	H	H	0.380	0.249	0.443	0.38
45	Cl	H	F	H	Me	H	0.544	0.636	0.680	0.411
46	NO2	H	Cl	H	Me	H	0.079	0.366	0.147	0.4
47	NO2	Me	F	H	Me	H	0.845	0.640	0.912	0.707
48	Cl	Cl	H	H	Me	H	1.690	1.475	1.459	1.396
49	Cl	H	H	H	OH	H	1.255	1.490	1.394	1.511
50	Cl	Me	H	H	OH	H	1.204	1.551	1.415	1.667
51	Cl	H	Cl	H	OH	H	0.544	0.746	0.303	0.698
52	Cl	H	F	H	Cl	H	0.724	0.676	0.761	0.688
53	H	H	F	F	H	Cl	1.447	1.257	1.388	1.476
54	H	H	F	H	H	Me	1.279	1.199	1.507	1.203
55	Cl	Me	H	H	H	Cl	1.602	1.519	1.416	1.136
56	Cl	H	F	H	H	Cl	0.556	0.796	0.621	0.569
57	Me	H	F	H	H	Cl	1.146	1.242	0.818	1.186

The QSAR models we obtained characterising benzdiazepine receptor inhibition significantly exceed in statistical characteristics the results of QSAR studies by the authors of the original paper [115] ($R<0.75$).

From Fig. 4.7.(a) shows that all information parameters are mainly concentrated in the substituent region and reflect the influence of each of them on the biological activity of 1,4-BD. The presence of bulk fragments in the substituent region R1 (Table 4.6.) (cells 92, 507) and R8 (cell 69), and electronegative fragments in the region of substituents R_2 (cell 4) will favour the increase of the inhibition parameter.

Characteristics of QSAR models describing the structure relationship of a series of 1, 4-benzdiazepin-2-ones with the benzdiazepine site inhibition parameter on the GABA family of$_A$ receptors

$$A = ax1 + bx_2 + cx_3 + dx_4 + ex_5 + fx_6 + g$$

Model characteristics	Model VII	Model VIII	Model IX
Number of selected parameters	6	6	6
a	-4.09	-0.48	-0.79
X1	DAW(92)	DAW(64)	DAW(67)
b	-0.78	-0.18	-0.009
X2	DvDW_Rad(507)	GGeom(6)	RotRf(70)
c	-0.52	-0.03	-0.04
X3	DAW(69)	DHyd(143)	DHyd(143)
d	0.27	2.39	0.002
X4	GL(6)	DGeom(5)	RotAw(733)
e	2.09	0.57	-0.28
X5	DAW(19)	DAW(44)	DAW(8)
f	-1.9	-3.31	0.82
X6	DEN(4)	DHyd(746)	D_R f(44)
g	8.89	-4.07	0.64
R	0.91	0.94	0.94
CVR	0.88	0.92	0.93
S	0.33	0.27	0.26
F	37	60	65

DAW(92), DAW(69), DAW(19), DAW(64), DAW(44), DAW(67), DAW(8) - divergence parameter of the atomic mass-weighted IP gradient field in cells 92, 69, 19, 64, 44, 67, 8, respectively; $_{DVDWRad}$(507) - divergence parameter of the gradient field of IP weighted by van der Waals radii of molecular atoms in cell 507;

G_L (6), G_{Geom} (6) is the gradient parameter of lipophilicity-weighted IPs and unweighted IPs in cell 6;

Rot_{Rf} (70) is the rotor of refraction-weighted VIPs in cell 70;

D_{Hyd} (143), D_{Hyd} (746) is the VIP divergence parameter of potential hydrogen bonds in cells 143, 746;

D_{Geom} (5) is the divergence parameter of the gradient field of unweighted FE gradient in cell 5;

Rot_{AW} (733) is the rotor of the VIP atomic mass distribution;

D_{EN} (4) is the divergence parameter of the electronegativity-weighted IP gradient field in cell 4;

D_R I(44) is the divergence parameter of the refraction-weighted IP gradient field in cell 44;

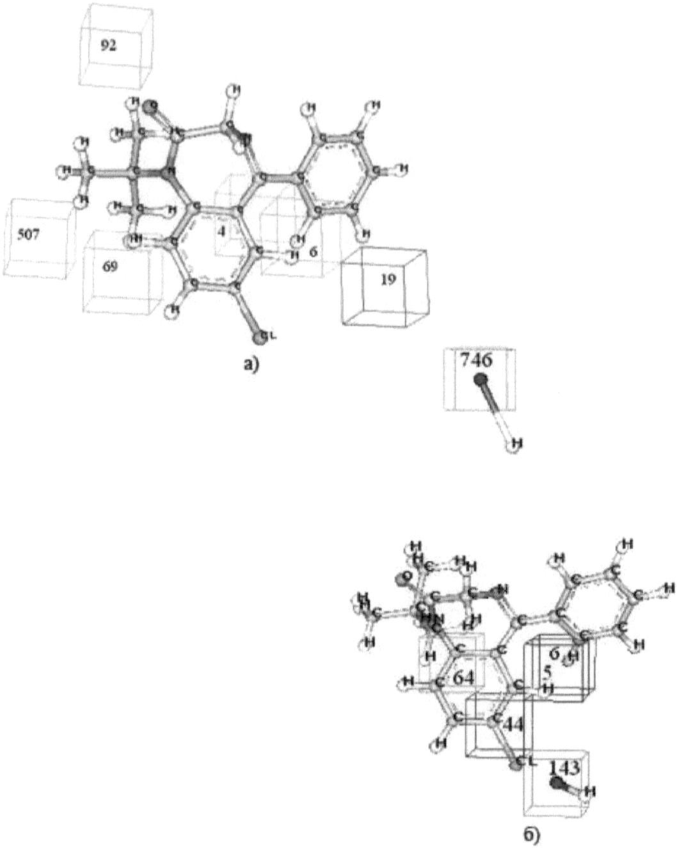

Fig.4.7. Cubic cells in which the information parameters of 1,4 - benzdiazepines determine the Log(IC_{50}) of the molecules: a) by model VII; b) by model VIII.

The negative influence on the inhibition parameter will be caused by the presence of bulk fragments in the region of substituents R_2 (cell 19).

In model VIII, as can be seen from Fig. 4.7. (b), all the IP parameters are located in the substituent region, except for the divergence parameter of vector IPs of potential hydrogen bonds (cell 746). The presence of bulk fragments in the region of substituents Ri (cell 64), as well as the presence of fragments capable of acting as hydrogen bond donors in the region of substituents R_7 (cell 143) will contribute to the increase of the inhibition parameter. It should be noted that the hydrogen bond vector IP parameter located outside the molecules in the region close to the R3 substituents (cell 746) also favours an increase in the inhibition parameter.

The presence of bulk groups in the region of substituents R_2 (cell 5) and R_7 (cell 44) will negatively affect the inhibition parameter of the investigated molecules.

In QSAR model IX (see Table 4.7. and Fig. 4.8), in addition to the scalar FE parameters and vector FE parameters of potential hydrogen bonds, vector FE parameters are also involved, taking into account the distribution of masses and refractions of atoms. As can be seen from Fig. 4.8, as in the previous models (VII and VIII), the presence of bulk groups (R1 substituents) (cell 67) will have a positive effect on the inhibition constant. In addition, the presence of bulky groups (substituents R_6) (cell 8), as well as the presence of fragments capable of being hydrogen bond donors in the region of substituents R_7 (cell 143) will also contribute to an increase in the inhibition parameter of the benzdiazepine site.

Thus, the developed concept of IP allows us to successfully solve and interpret problems related to the study of the influence of the structure of compounds on their biological activity, i.e. information parameters can be successfully used as structural descriptors for predicting various properties of molecules.

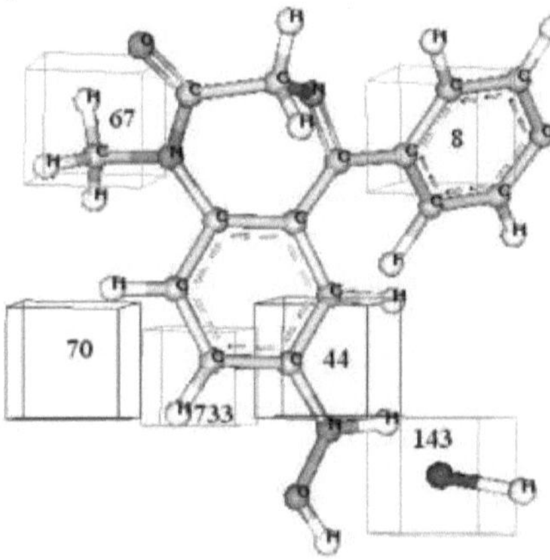

Figure 4.8: Cubic cells in which the information field parameters of 1,4-benzdiazepines determine the $Log(IC_{50})$ of the molecules (Model IX).

4.3 Analysis of the influence of the structure of lanthanide (III) P-diketonates on the luminescence intensity

In addition to the structure-bioactivity tasks, in the present work we have tried to analyse the effectiveness of the developed approach to the study of other properties.

The continuing interest of researchers in the study of the structure and properties of |3-lanthanide diketonates (Ln) is caused by their unique properties, due to which they are widely used in the creation of light-transforming and other optical materials, as shift reagents in NMR spectroscopy, luminescent probe in immunofluorescence analysis, etc. [116-119]. [116-119].

The largest number of works refers to the study of the properties of Ln compounds with |3- diketones (mainly acetylacetone AA derivatives) containing short alkyl radicals. The structure of these compounds in the solid phase was studied by X-ray diffraction [118], and in solutions - by NMR spectroscopy [120], which allows us to determine not only the number of ligands, but also the way of their coordination and the chemical structure of the complex. The molecular mechanics method was applied by the authors [121] to calculate the spatial structure of acetyl- and benzoylacetonates Ln - $Ln(AA)_3$, Ln(EA) .3

As is known [119], one of the factors limiting the possibility of observing the intense luminescence of Ln in aqueous solutions of their complexes is its quenching by vibrations of OH-groups of water molecules (OH-oscillators) included in the inner coordination sphere of the complex. Its complete or partial displacement occurs upon introduction of a second ligand, usually an organic base. The study of optical characteristics of Ln complexes with phenyl and a- thienyl acetylacetone derivatives containing different fluoroalkyl substituents has shown the presence of a correlation between the intensity, quantum yield, and luminescence lifetime of Ln P-diketonates and the chain length of the fluorinated radical (RF) [122-124]. It was found [125] that elongation of R_F FROM CF_3 TO C F_{613} leads to complete displacement of water molecules from the inner coordination sphere of the complex without introduction of a second ligand at R_F = C F_{613} .

The fact that C F_{613} is a residue of perfluoroenanthic acid used as a surfactant made it possible to put forward a hypothesis about the formation by long fluorocarbon radicals (C F_{613} -C F_{817}) of a highly hydrophobic shell capable of shielding the central ion from the quenching action of water molecules [126]. The propeller-shaped arrangement of chelate rings in the complexes of Ln with perfluoro-P diketones was established by molecular mechanics and NMR spectroscopy. The long fluorocarbon radicals (C Fi_{63}) were found to form a close to spherical shell around the central ion. The calculated values of hydrophobicity (logP) of fluorinated P-diketones are 2-3 times higher than the values of this parameter for non-fluorinated analogues. Thus, creating a highly hydrophobic shell around the Ln ion, ligands with R_r =C F_{613} -C F_{817} prevent the penetration of water molecules into the inner coordination sphere of the complex, which determines the high luminescence intensity of perfluorinated P-diketonates Ln.

In this work, we have attempted, using the IP concept, to establish a relationship between the luminescence intensity of [3-diketonates of Ln and lipophilicity-weighted IP parameters [127].

In this work, phenyl and a-thienyl derivatives of AA with different chain lengths of the fluorinated radical (R_r =CF_3 -C F_{817}), including those with the inclusion of an oxygen atom (CF_3 OCF_2 -, CF_3 OCF_2 CF2-) were analysed and AA and BA were compared.

The tris-chelate |3-diketonate complexes Sm^{3+} , Eu^{3+} , Nd^{3+} , Yb^{3+} (Table 4.8.) were considered as objects of study. Since under identical conditions the luminescence intensity of [3-diketonates Eu^{3+} (calculated) is an order of magnitude higher than that of Sm^{3+} , 2 and more orders of magnitude higher than that of Yb^{3+} and Nd^{3+} , its registration was carried out under different conditions (Ln concentration^{3+} and spectrometer slit opening), so it is unfortunately not possible to solve the structure-property problem simultaneously for compounds Sm^{3+} , Eu^{3+} , Nd^{3+} and Yb^{3+} . In this connection, the corresponding regression models were constructed for four samples, each of which included complexes of one of the studied metals (Table 4.9.).

The compounds of this series were considered within the framework of the molecular lattice model [109]. At the initial stage of work, the molecules of the investigated series were optimised using the universal force field (UMF) [128]. Next, a conformational search was performed for all metal complexes, resulting in a set of conformers for each molecule individually. From these sets we selected the most energetically favourable conformations.

The obtained conformations were matched by metal atoms and associated oxygen atoms. Then, after such operations, lattice models were constructed for each molecule and more than 3 thousand structural parameters were calculated. In each lattice cell, the potential, gradient and divergence of the gradient field of the information fields weighted by the lipophilicity of the molecule atoms were calculated. After screening out mutually correlated parameters, models with quite satisfactory statistical characteristics were obtained for each subsample by the stepwise regression method (Table 4.9.), where D_{Lip} *(890)*, D_{Lip} *(590)*, D_{Lip} *(913)* are divergences of lipophilicity-weighted IP gradient fields in cells 890, 590, and 913, respectively.

Table 4.8.

Luminescence intensity ($1lu_M$) of Sm, Eu, Nd, and Yb complexes with acetylacetone derivatives R1 - COCH2CO - R2.

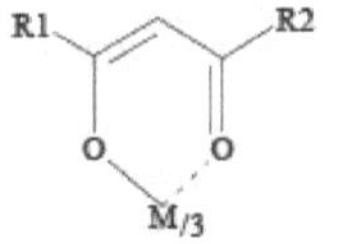

M : Eu, Nd, Sm, Yb

No.	P-diketones (substituents)		[люм ., OTH. unit.			
n/a	R1	R2	Sm	Eu	Nd	Yb
1	SNz -	CH3 -	1	2	3	3
2	SNz -	C6H5 -	7	14	15	11
3	CF3 -	C6H5 -	38	43	35	45
4	C3F7 -	C6H5 -	55	80	41	59
5	C6F13 -.	C6H5 -	110	150	98	93
6	C8F17 ˙	C6H5 -	130	206	110	98
7	CF3-O-CF2 -	C6H5 -	60	85	45	54
8	CF3-O-CF2-CF2 -	C6H5 -	68	101	66	78
9	C3F7-O-CF(CF3) -	C6H5 -	70	162	60	62
10	CF3-O-(CF2)3 - O - CF(CF3) - -	C6H5 -	95	181	82	84
11	CF3 -	C4H3S[10] -	14	18	30	37
12	C3F7 -	C4H3S -	45	62	38	53
13	C6F13 -.	C4H3S -	89	120	78	84
14	CF3-O-(CF2)3 - O - CF(CF3) - -	C4H3S -	61	75	43	51
15	CF3-O-CF2 -	C4H3S -	90	174	75	79

Fig. 4.9 shows the regions of space in which the information parameters have a decisive influence on the luminescence of lanthanide complexes. As can be seen from the figure, these parameters are located in the region of fluoroalkyl groups.

Thus, the obtained results are fully consistent with the hypothesis about the creation by fluoroalkyl groups of a hydrophobic barrier preventing the penetration of water molecules into the inner coordination sphere of the complex.

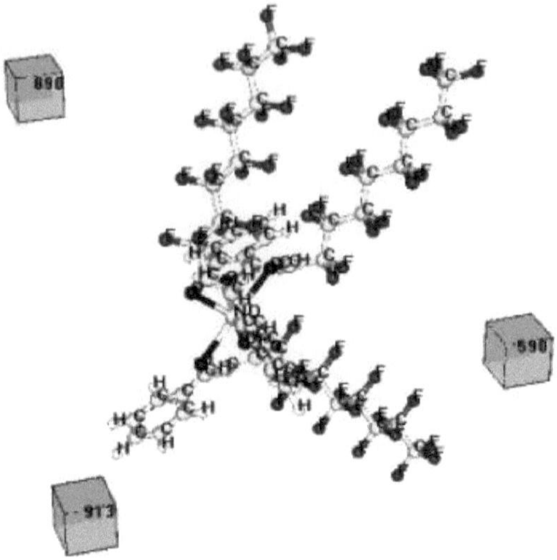

Figure 4.9. Regions of space (cells 590, 913 and 890) in which the divergence parameters of lipophilicity-weighted IPs have a decisive influence on the luminescence of lanthanide complexes.

Table 4.9.

Regression models of the structure-luminescence intensity relationship
Y=A+B-X **for lanthanide complexes with β- -diketones.**[10]

Y	A	B	X	R	CVR	S	F
1lum ( )^{Eu}	-179.2	3714.6	*DLip(890)*	0.95	0.94	21.6	114.6
1lum (Nd)	-33.6	1132.9	*DLip(590)*	0.95	0.94	9.4	130.4
1lum $^{(Sm)}$	-51.7	1441.7	*DLip(590)*	0.94	0.93	13.1	101.7
1lum $^{(Yb)}$	-12.4	1155.0	*DLip(913)*	0.91	0.88	10.5	85.4

As can be seen from Table 4.9., the quality of the regression equations is quite good satisfactory - correlation coefficients R = 0.91 - 0.95, correlation coefficients obtained under sliding control conditions (CVR = 0.88 - 0.94) are very high. The comparison of calculated and experimental data is presented in Fig. 4.10.

1010 α -thienyl

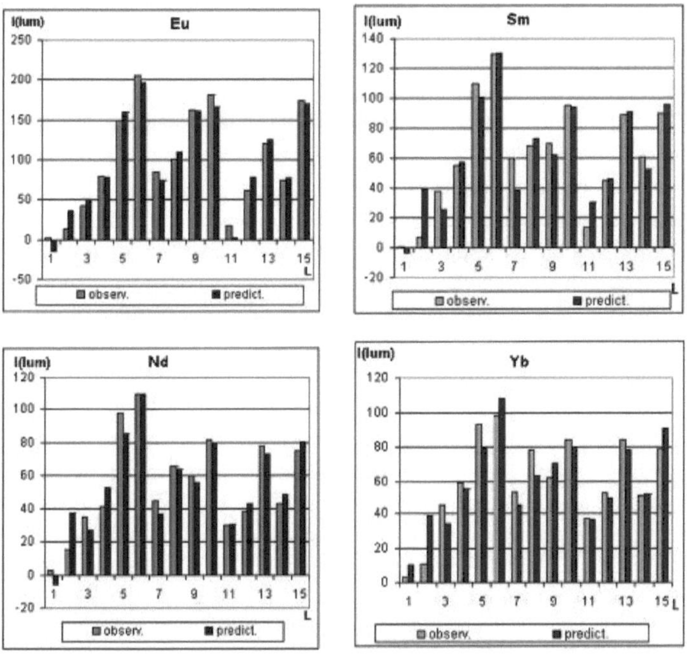

Fig. 4.10. Initial and calculated values of luminescence intensity of the complexes Ln.

Thus, the obtained structure-property models are quite suitable for preliminary screening of the luminescence intensity of [3-diketonate Ln complexes. In all probability, expansion of the training sample by including new complexes will improve the quality of the corresponding structure-property models in the future.

Conclusions

The main results of the present work are the creation of the IP concept, systematic analysis of the possibilities of its application for solving "structure-property" problems, and identification of the regularities of the influencc of various IP parameters of molecules on their properties.

1. The concept of a potential information field, from a formal point of view, reflects the situation when an object (molecule) structures the surrounding space, i.e. generates information in it. Models realising this concept make it possible to construct scalar and vector information fields weighted by various properties (mass, charge, lipophilicity, etc.). Such fields, in fact, characterise information about the distribution of the considered property in space.
2. In a number of model systems (various benzene derivatives) it is shown that at systematic variation of molecular structure various parameters of the corresponding information fields change in a regular way, which allows us to use the parameters of molecular information fields for solving various "structure-property" problems.
3. The model of the molecular information field of chirality reflects the ability of a chiral molecule to dissymmetrize the surrounding space. It was found that only for achiral molecules the positive ("right") and negative ("left") regions of the chirality information field potential fully compensate each other. For enantiomers of molecules with axial symmetry, the potential of the chirality information field has a constant sign: (+) for "right" and (-) for "left" stereoisomers.
4. On model systems it is shown that for supramolecular compounds the complementarity of the information fields of the "host" and "guest" molecules, estimated on the basis of correlation and autocorrelation characteristics of these fields, is maximal for pairs of molecules forming the most stable complexes.
5. The sequence of non-classical bioisosters, ordered in terms of their structural similarity, unlike classical bioisosters, depends to a significant extent on weight parameters (atomic characteristics), on the basis of which the corresponding information fields are constructed.
6. The informational influence of a-amino acids on the environment correlates with their essentiality, reflecting the important informational role of a-amino acids in the early stages of abiogenesis.
7. The use of only the parameters of various information fields as structural characteristics of molecules allowed us to effectively solve a variety of "structure-property" problems. It was shown that an important factor influencing the binding of steroids to corticosteroid- and testosterone-binding globulin is the possible formation of intermolecular hydrogen bonds of oxygen-containing fragments; the most adequate models describing the interaction of 1,4-benzdiazepin-2-ones with the benzdiazepine site in the family of GABAcd receptors were constructed on the basis of parameters of both scalar and vector information fields, taking into account the distribution of masses and refractions of atoms; the luminescence intensity of P-diketonate REE complexes can be estimated on the basis of the characteristics of the potential, gradient, and divergence of the gradient field of the corresponding scalar information fields weighted by lipophilicity.

REFERENCE LIST

1. *Amosov N.M.* Modelling of complex systems. - K.: Naukova Dumka, 1968. - 88 c.
2. *Blokh I.I.* Basic concepts of information theory. - L.: Znanie, 1959. - 28 c.
3. *Zarenin Yu.G.* Corrective codes for information transmission and processing. - K.: Technics, 1965. - 170 c.
4. *Kochergin A.N.* Modelling of thinking. - Moscow: Gospolitizdat, 1969. - 224 c.
5. *Lerner A. Ya.* Beginnings of Cybernetics. - Moscow: Nauka, 1967. - 400 c.
6. *Wilson R.* Optical reading devices: Per.s Engl. - M.: Mir, 1969. - 216 c.
7. *Brillouin L.* Science and Information Theory: Per. from Engl. - M.: Fizmatgiz, 1960. - 392 c.
8. *Blushkov V.M.* Thinking and Cybernetics // Questions of Philosophy. - 1963. - №1. - C. 36 - 48.
9. *Becker L.M.* Perception and the basis of its modelling. - L.: Nauka, 1964. - 194 c.

10.*Zal'tsberg B.* What is Information // The Concept of Information and Biological Systems. - M.: Mir, 1963. - C. 13 - 23.

11 *Bokarev V.A.* Cybernetics and Military Affairs. - Moscow: Voenizdat, 1969. - 287 c.

12 *Petrovsky A.V.* General Psychology. - M.: Prosveshchenie, 1970. - 132 c.

13 .Petrushenko *L.A.* Self-movement of matter in the light of cybernetics. - M.: Nauka, 1971. - 292 c.

14 . *Seravin L.N.* Information theory from the biologist's point of view. - L.: LGU, 1973. - 160 c.

15 .*Vasilieva T.P. The* Laws of Information. - M.: Amrita - Rus, 2002. - 320 c.

16 *Shannon K.* Works on information theory and cybernetics: Per. from English - M.: IL, 1963. - 829 c.

17 .*Brillouin L.* Scientific uncertainty and information. - M.: Mir, 1966. - 498 c.

18 .*Castler B.* The Emergence of Biological Organisation. - M.: Mir, 1967. - 90 c.

19 . *Thomas D. Schneider* A Glossary for Molecular Information Theory. *http://www. lecb. ncifcrf.gov/thoms/glossary. html.*

20 . *Theory of* information and its applications. - Moscow: Fizmatgiz, 1969. - 467 c.

21 *Yaglom A.M., Yaglom I.M.* Probability and information. - M.: Nauka, 1973. - 511 c.

22 *Chemical* Applications of Topology and Graph Theory: Translated from English - M.: Mir, 1987. - 560 c.

23 *Shambabal P.* Development and application of the entropy concept. - M.: Nauka, 1967. - 245 c.

24 *Kolmogorov A.N.* Three approaches to the definition of the concept of "quantity of information" // Problems of information transfer. - M., 1965. - vol. 1, issue 1. - 534 c.

25 *Kolmogorov A.N.* To logical bases of information theory and probability theory // Problems of information transfer. - M., 1969. - vol. 5, issue 3. - 564 p.

26 *Ashby W.* Introduction to Cybernetics. - Moscow: Izd-wo foreign lit., 1959. - 351 c.

27 *Glushkov V.M.* Cybernetics, Thinking, Life. - Moscow: Mysl, 1964. - 326 c.

28 *Harari F.* Graph Theory. - M.: Mir, 1973. - 302 c.

29 . *Wilson R.* Introduction to graph theory. - M.: Mir, 1977. - 427 c.

30 *Dmitriev I.S.* Molecules without chemical bonds. - L.: Khimiya, 1980. - 382 c.

31 *Zhbanov Y.A.* Entropy of information in organic chemistry. - Rostov-on-onu:Izd-vo RSU, 1979. - 56.
32 *Zhbanov Y.A.* Theory of structure of organic compounds. - Moscow: Higher School, 1971. - 295 c.
33 . *Artificial* Intelligence: applications in chemistry / Edited by T. Pierce and B. Honey: Transl. from Engl. - M.: Mir, 1988. - 430 c.
34 .*Mazur M.* Qualitative theory of information. - M.: Mir, 1974. - 268 c.
35 .*Wiener N.* Cybernetics. - Moscow: Soviet Radio, 1968. - 194 c.
36 .*Shu-Kun Lin.* Correlation of Entropy with Similarity and Symmetry // J. Chem. Chem. Inf. Comput.Sci. - 1996. - V. 36. - P. 367 - 376.
37.*Dimitris K. Agrafiotis.* Stochastic Algorithms for Maximizing Molecular Diversity // J. Chem. Chem. Inf. Comput. Sci. - 1997. - V. 37. - P. 841 - 851.
38.*Klimontovich Yu.L.* Criteria of the relative degree of orderliness of open systems // Uspekhi fiz. nauki. - 1996. - т. 166, № 11.
39.*Orismology* (the Science of Defining Words) and the Geometrical Foundations of Chemistry. 5. The Heuristics of Primacy // J. Chem. Chem. Inf. Comput. Sci. - 1997. - V. 37. - P. 835 - 840.
40 .*Gasteiger J., Sadowski J., Schuur J., Selzer P., Steinhauer L., and Steinhauer V.* Chemical Information in 3D Space // J. Chem. Inf. Comput. Sci. - 1996. - V. 36. - P. 1030 - 1037.
41 .*Dimitris K. Agrafiotis*. On the Use of Informational Theory for Assessing Molecular Diversity // J. Chem. Chem. Inf. Comput. Sci. - 1997. - V. 37. - P. 576 - 580.
42 .*Daniel J. Graham.* Information and Organic Molecules: Structure Considerations via Integer Statistics // J. Chem. Chem. Inf. Comput. Sci. - 2002. - V. 42. - P. 215 - 221.
43 *Shmal'gauzen I.I.* Cybernetic questions of biology. - Novosibirsk: Nauka, 1968. - 354 c.
44 *Volkenstein M.V.* Physics and Biology. - M.: Nauka, 1980. - 152 c.
45 *Koptiug V.A.* State and prospects of using computers for solving problems of organic chemistry // Zh.-l VHO. - 1985. - т. 30, № 1. - C. 102 - 105.
46 .*Gombar V.K., Wadhwa L.* Quantitative structure-activity relationships. IV. Use of molecular negentropy as structural parameter // Arzneimittelforschung. - 1982. - V. 32, Issue 7. http ://reviews.bmn. com/medline.
47 .*Kier L.B.* Use of molecular negentropy to encode structure governing biological activity // J. Pharm. Pharm. Sci. - 1980. - V. 69, Issue 7. http ://reviews .bmn. com/medline.
48 .*Godden J.W., Bajorath J.* Chemical descriptors with distinct levels of information content and varying sensitivity to deferences between selected compound databases identified by SE-DES analysis // J. Chem. Chem. Inf. Comput. Sci. - 2002. - V. 42, Issue 1. http://reviews.bmn.com/medline.
49 .*Strait B.J., Dewey T.G.* The Shannon information entropy of protein sequences. Biophys. J. - 1996. - V. 71, Issue 1. http ://reviews.bmn. com/medline.
50 .*Wertz S.* Mathematical model of molecular complexity // Chemical applications of topology and graph theory. - M.: Mir, 1987. - C. 236 - 258.
51. *Magnuson V., Harris D, Beysak S.* Topological indices based on neighbourhood

symmetry: chemical and biochemical applications // Chemical applications of topology and graph theory. - M., "Mir", 1987. - C. 206 - 221.
52. *Rudkevich M.I., Rudkevich D.M., Pyrin A.I.* Application of information theory to coal tar chemistry // Coke and Chemistry, 1986. - №4. - C.30 - 32.
53. *Dembovskiy S.A., Kozmin P.A.* Order, disorder and geometrical information indices of molecules // Izv. of AN. Ser. of chemical sciences. - 1996, № 8. - C. 1912 - 1926.
54. *Kozmin, P.A.* Quantitative estimation of order in the spatial arrangement of atoms in a solid // Reports of the Academy of Sciences. - 1995, Vol. 343, No. 2, - P. 205 - 206.
55. *Kozmin, P.A.* Quantitative estimation of the order in the spatial arrangement of atoms in crystals and amorphous bodies // Inorganic Materials. - 1995. - т. 31, № 7. - C. 884 - 887.
56. *Bonchev DG.* Characterisation of chemical structures by means of information theory and graph theory: Cand. Dr. of chemical sciences, Moscow State University, Moscow - 1984. - 409 c.
57. *Bonchev D.* Information Theoretic Indices for Characterisation of Chemical Structures // Research Studies Press, Chichester. - 1983. - P. 249.
58. *Kuz'min V.E., Ognichenko L.N., Artemenko A.G.* Modeling of the informational field of molecules // J.Mol.Model. - 2001. - № 7. - P. 278 - 285.
59. *Kuzmin B.E., Artemenko A.G., Onishchenko S.V., Ognichenko LN.* Modelling of the information field of molecules // Theses of reports of the 1st All-Russian Conference "Molecular Modelling". - Moscow, 1998. - У 7.
60. *A.G. Artemenko, V.E. Kuz'min, L.A. Ognichenko, S.J. Kuz'min, L.A. Ognichenko, S.J. Onischenko. Onischenko.* New Informational - Field Characteristics of Molecular Recognition for Problem Solving "Structure - Property" // 7th International Conference of Mathematical Chemistry. - Girona (Spain), 1997. - P. 100.
61. *V.E. Kuz'min, A.G. Artemenko, L.N. Ognichenko Modeling of the informational field of molecules // Dutch - Ukrainian International Colloquium on Catalysis. Ognichenko* Modeling of the informational field of molecules // Dutch - Ukrainian International Colloquium on Catalysis. - Kiev, 2000. - P. 20.
62. *Onishchenko C.B., Ognichenko LN* Modelling of the information field of molecules // Proc. of the 1st conference of young scientists of the Southern region of Ukraine. 1^{ои} conf. of young scientists of the Southern region of Ukraine. - Odessa, 1998. - C. 33.
63. *Ogtchenko L.M.* Transformation of structural parameters of molcules on the basis of! ix shformatsshnyh po.pv // Proc. of 3ой conf. of young scientists of tvdennogo reponu Ukraugnya. - Odessa, 2000. - C. 21.
64. *Ognichenko L.N., Artemenko A.G.* Modelling of the information field of a molecule // Proc. of the Third! All-Ukrainian Conference of Young Scientists Information Technologies in Science, Education and Technology (ITOHT - 2002). - Cherkasy, 2002. - C. 75 - 78.
65. *Kuzmin V.E., Ognichenko L.N., Artemenko A.G., Lyakhovsky A.L.* Applied possibilities of the information field of a molecule // Proc. of the 3rd All-Russian Conference "Molecular Modelling". 3^{01} All-Russian Conf. "Molecular Modelling". - Odessa, 2003. - C. 80.
66. *Ognichenko L.N., Kuzmin V.E.* Natural-scientific conception of information field of

material object // Theses of the seminar "Information systems and technologies". - Odessa, 2003. - C. 42.
67. *Kuzmin V.E., Ognichenko L.N.* Applied aspects of the concept of the information field of a molecule // Abstracts of the seminar "Information Systems and Technologies". - Odessa, 2003. - C. 41.
68. *Ursul A.D.* Reflection and Information. - Moscow: "Nauka", 1973. - 135 c.
69. *Goldfein I.A.* Vector analysis and field theory. - Moscow: Fizmatgiz, 1962. - 132 c.
70. *Ignatsius G.I.* Field Theory. - Moscow: Znanie, 1971. - 112 c.
71. *Ogtchenko L.M., Chelombit'ko V.A., Otschenko S.V.* Molecular formazshne field of ch1ralnost1 // Proc. of the 5th conference of young scientists and students. 5^{01} conf. of young scientists and students in the field of topical nativities. - Dnshropetrovsk, 2003. - C. 34.
72. *Kuz'min V.E., Chelombitko V.A., Yudanova I.V., Stelmakh I.B., Rublev I.S.* Stereochemical analysis by simplex representation of molecules // J.Struct.Chem. - 1998. - V. 39, No. 3. - P. 549 - 554.
73. *Alikhanidi S.E., Kuz 'min V.E.* Quantitative evaluation of molecular chirality on the base of an optimised approach to comparison of original and reflected structures // Dopov. HAH Ukragni. - 1999. - № 3. - C. 138 - 141.
74. *Kuzmin V.E.* Modelling of the information field of molecules. // Dokl. NAS of Ukraine. - 2000. - № 3. - C. 159 - 163.
75. *Potapov V.M.* Fundamentals of Stereochemistry. - M.: Mir, 1972. - 289 c.
76. *Ognichenko L.N., Kuz'min V.E., Chelombitko V.A.* The Analysis of Molecular "Guest - Host" Complexes Recognition within Molecular Field Concept // Second International Symposium "Molecular Design and Synthesis of Supramolecular Architectures". - Kazan (Russia), 2002. - P. 68.
77. *Ognichenko L.N., Kuzmin B.E., Chelombit'ko B.A.* Analysis of molecular recognition for "guest-host" complexes within the concept of information field // Ukrainian-Polish-Moldavian Symposium on supramolecular chemistry. - Kiev, 2003. - C. 173 - 175.
78. *Hyperchem* software. Hypercube, Inc, 1115 NW 4th Street, Gainesville, FL 32601 USA. http://www.hyper.com.
79. *Zefirova O.H., Zefirov NS.* On the history of the origin and development of the concept of bioisosterism. // Vesti. Moscow University. Ser.2. Chemistry. - 2002. - T. 43, № 4. - C. 251 - 256.
80. *Hansch C., Fujita T.* A method for the correlation of biological activity and chemical structure // J. Am. Am. Chem. Soc. - 1964. - V. 86, № 8. - P. 1616 - 1626.
81. *Ognichenko L.N.* Quantitative assessment of isosterism (bioisosterism) of organic compounds based on the concept of information field // Proc. of the International Conference of Students and Postgraduates "Modern directions of chemistry development". - Odessa, 2004. - C. 34.
82. *Davison M.* Multidimensional scaling. Methods of visual representation of data. - M.: Finance and Statistics, 1988. - 254 c.
83. *Ognichenko L.N., Kuzmin V.E., Muratov N.N.* Analysis of the information fields of amino acids // Proc. of Odessa *Polytechnic* University. Odessa Polytechnic University. - Odessa, 2002. Vol. 2. - C. 228 - 232.

84. *Mekler L.B.* On the Origin of Living Cells: Evolution of Biologically Significant Molecules - Transition of Chemical Evolution to Biological Evolution. A New Approach to the Problem // J. All-Union Chemical Society named after D.I.Mendeleev. - 1980. - т. 25, № 4. - С. 460 - 473.
85. *Morgan H.L.*. The Generation of a Unique Machine Description for Chemical Structures-A Technique Developed at Chemical Abstracts Service // J. Chem. Chem. Doc. - 1965. - 5 (2). - P. 107 - 113.
86. *Shelley C. A., MunkM. E.* Computer Perception of Topological Symmetry // J. Chem. Chem. Inf. Comput. Sci. - 1977. - 17 (2). - P. 110 - 113.
87. *Shelley C. A., MunkM. E.* An Approach to the Assignment of Canonical Connection Tables and Topological Symmetry Perception// J. Chem. Chem. Inf. Comput. Sci. - 1979. - 19 (4). - P. 247 - 250.
88. *Uchino M.* Algorithms for Unique and Unambiguous Coding and Symmetry Perception of Molecular Structure Diagram. I. Vector Function for Automorphism Partitioning// J. Chem. Chem. Inf. Comput. Sci. - 1980. - 20 (2). - P. 116 - 120.
89. *Golender V. E., Drboglav V. V., andRosenblit A. B.* Graph potentials method and its application for chemical information processing // J. Chem. Chem. Inf. Comput. Sci. - 1981. - 21 (4). - P. 196 - 204.
90. *Ruecker G., Ruecker Ch.* Computer perception of constitutional (topological) symmetry: TOPSYM, a fast algorithm for partitioning atoms and pairwise relations among atoms into equivalence classes // J. Chem. Chem. Inf. Comput. Sci. - 1990. - 30 (2). - P. 187 - 191.
91. *Liu Xiaoyu, Balasubramanian K., Munk M. E.* Computational techniques for vertex partitioning of graphs // J. Chem. Chem. Inf. Comput. Sci. - 1990. - 30 (3). - P. 263 - 269.
92. *Figueras J.* Automorphism and equivalence classes// J. Chem. Chem. Inf. Comput. Sci. - 1992. - 32 (2). - P. 153 - 157.
93. *Razinger M.* Extended Connectivity in Chemical Graphs// Theoret. Chim. Acta (Berl.). - 1982. - 61. -P. 581 - 586.
94. *Skorobogatov B.A., Khvorostov P.V.* Collected in: Algorithms for Structural Information Analysis. (Computing systems, 103) / Ed. V.A. Skorobogatov. - Novosibirsk, 1984. - С. 6-26.
95. *Gitlina L.S., Golender V.E., Drboglav V.V. et al.* Methods of Representation and Processing of Structural Information for Analysis of Structure-Activity Relationship. - Riga, 1981.
96. *Fowler P.W., Manolopoulos D.E.*. An Atlas of Fullerenes. - Oxford: Clarendon Press, 1995.
97. *Kuzmin B.E., Zheltvai A.I., Chelombit'ko B.A., Alikhanidi S.E.* Stereoanalysis of fullerenes with chiral molecular backbone // J. Structural Chemistry. - 2005. - T.46, № 5. - С. 945 - 952.
98. *Sukhachev D.V., Pivina T.S., Shlyapochnikov V.A., Petrov E.A., Palulin V.A., Zefirov N.S.* Investigation of quantitative relations "structure - sensitivity to impact" of organic polyazotized substances // Dokl. RAS. - 1993. - т. 328, № 2. - С. 50 - 51.
99. *CODESSA* PRO User's Manual // University of Florida, 2003.
100. *Seel M., Turner D.B., Wilett P.* HQSAR - a highly predictive QSAR technique

based on molecular holograms // QSAR. - 1999. - V. 18. - P. 245 - 252.
101. *Hurst T., Heritage T.* HQSAR - A Highly Predictive QSAR Technique Based on Molecular Holograms // 213th ACS Natl. Meeting, San Francisco, CA. - 1997. - P. 19.
102. *Cramer R.D., Patterson D.E., Bunce J.D.* Comparative Molecular Field Analysis (CoMFA). 1.Effect of Shape Binding to Carrier Proteins // J.Am.Chem.Soc. - 1988. - V. 110, № 18. - P. 5959 - 5967.
103. *Merz A., Rognan D.* CoMFA: Scope and Limitations. In 3D QSAR in Drug Design / Ed. H.Kubinyi . - ESCOM: Leiden, 1993. - P. 583 - 618.
104. *Förster E., Röntz B.* Methods of correlation and regression analysis // M.: Finance and Statistics, 1983. - 244 c.
105. *Mathematical* Statistics / V.M. Ivanova, V.N. Kalinina, L.A. Neshumova, I.O. Reshetnikova; Edited by A.M. Dlin. - Moscow: Higher School, 1975. - 398 c.
106. *Good A.C., Sung-Sau So, Richards W.G.* Structure - Activity Relationships from Molecular Similarity Matrices // J.Med.Chem. - 1993. - V. 36. - P. 433 - 438.
107. *Good A.C., Peterson J.S., Richards W.G.* QSAR's from Similarity Matrices. Technique Validation and Application in the Comparison of Different Similarity Evaluation Methods // J.Med.Chem. - 1993. - V. 36. - P. 2929 - 2937.
108. *Artemenko A.G.* Analysis of structure-property relationship on the basis of molecular lattice models. Dne. ... Cand. of chemical sciences: 02.00.03. - Odessa, 2000. - 151 c.
109. *Kuz'min V.E., Artemenko A.G, Kovdienko N.A., Tetko I.V., Livingstone D.J.*. Lattice Model for QSAR Studies // J. Mol. Mol. Modeling. - 2000. - № 6. - P. 517 - 526.
110. *Burkert W., Ellinger H* Molecular Mechanics: Per. with Engl. - M.: Mir, 1986.
111. *Kuz'mt, V.S.; Ogtchenko, L.M.* Modelling of the waterborne water communication data field and its application for QSAR problem solving // Vknik of Odessa National University. (issue "Ch1m1ya"). - 2003. - T. 8, № 7. - C. 117 - 123.
112. *Ogtchenko L.M.* Model of the formatzsch field of water links i 11 application for solving QSAR problems // Proc. of the 5th conference of young scientists of the Ukrainian State University of Regions // Proc. of the 5th conference of young scientists of the Ukrainian State University of Regions. 5ой conf. of young scientists of the Ukrainian region. - Odessa, 2002. - C. 48.
113. *Johansen A. V.* The rule of product of acid-base functions of molecules at their association by hydrogen bonds in solutions in CC14 // Theoret. and experimental chemistry. - 1971. - Vol. 7. 3. - C. 302 - 311.
114. *Kuzmin VE, Artemenko AG, Ognichenko LN, Voliansky VE, Gotsulyak LE, Sidelnikova TA, Kuzmina AV A* new approach to predicting the biological activity of derivatives of 1,4-benzdiazepines as components in the complex therapy of stroke // Vest. mor. med. - 2002. - № 3. - C. 47 - 51.
115. *Winkler D.A., Frank R.B., Watkins A.J.R.*. Atomistic Topological Indices Applied to Benzodiazepines using Various Regression Methods // Quant. Struct.-Act. Relat. - 1998. - V. 17. - P. 14 - 19.
116. *D Diketonates of* metals / Edited by Spitsyn V.I. - M.: Nauka, 1978. - 123 c.
117. *Problems of* chemistry and application of metal P-diketonates / Edited by Spitsyn V.I. - M.: Nauka, 1982. - 264 c.

118. *Theoretical* and applied chemistry of metal P-diketonates / Edited by Spitsyn V.I. and Martynenko L.I. - M.: Nauka, 1985. - 271 c.

119. *Poluektov N.S., Kononenko L.I., Efryushina N.P., Beltyukova S.V.* Spectrophotometric and luminescent methods for determination of lanthanides. - K.: Naukova Dumka, 1989. - 256 c.

120. *Marov I.N., Kostromina N.A.* EPR and NMR in the chemistry of coordination compounds. - Moscow: Nauka, 1979. - 268 c.

121. *Dubrovsky T.B., Chudinov A.V., Savitsky A.P., Timofeeva TV, Struchkov Y.T.* Use of molecular mechanics to calculate the molecular structure of eight-coordinated complexes of europium // Izv. RAS Sciences. Ser. chem. 1993. - № 9. - C. 1555 - 1559.

122. *Rusakova N.V., Topilova Z.M., Meshkova S.B.* Relation of stability and spectral-luminescent characteristics of P-diketonates of neodymium, samarium and europium with the properties of ligands // Zhury. neorg. chemii. - 1992. - T.37, № 1. - C. 116 - 120.

123. *Rusakova N.V., Meshkova S.B., Venchikov V.Ya.* On the relationship of quantum yield and luminescence duration of neodymium P-diketonate complexes with properties of ligands // Jury. of Applied Spectroscopy. - 1992. - T. 56, № 5 - 6. - C. 799 - 804.

124. *Topilova Z.M., Meshkova S.B., Bolshoi D.V.* Infrared luminescence of ytterbium ions in complexes with fluorine-containing P-diketones // Zhury. neorganic chemistry. - 1997. - T. 42, № 1. - C. 99 - 104.

125. *Meshkova S.B., Topilova Z.M., Lozinsky M.O.* Quenching of lanthanide luminescence in complexes with P-diketones containing various fluorinated radicals // Zhury. prikl. spectroscopii. - 1997. - T. 64, № 2. - C. 217 - 221.

126. *Meshkova S.B., Shapiro Y.E., Kuzmin V.E.* Conformational analysis of lanthanum(SH) and neodymium(SH) complexes with perfluoro-P-diketones // Coord. chem. - 1998. - T. 24, № 9. - C. 714 - 718.

127. *V.E. Kuz'min, L.N. Ognichenko, A.G. Artemenko, S.B. Meshkova. Meshkova.* Relationships of a luminescent emission of lanthanide complexes of P-diketones with parameters of an informational field of these molecules // Abstracts of XX Intern. Chugaev Conf. on "Coordination Chemistry". - Rostov-on-Don, 2001. - C. 543 - 544.

128. *Rappe A.K., Casewit C.J., Colwell K.S., Goddart W.A., Skiff W.M.* UFF, a Full Periodic Table Force Field for Molecular Mechanics and Molecular Dynamics Simulations // J. Am.Chem. Am.Chem. Soc. - 1992. - V.114. - P. 10024 - 10035.

APPENDICES.

APPENDIX A

Parameter maps of the information fields of the host (h) and guest (g) molecules

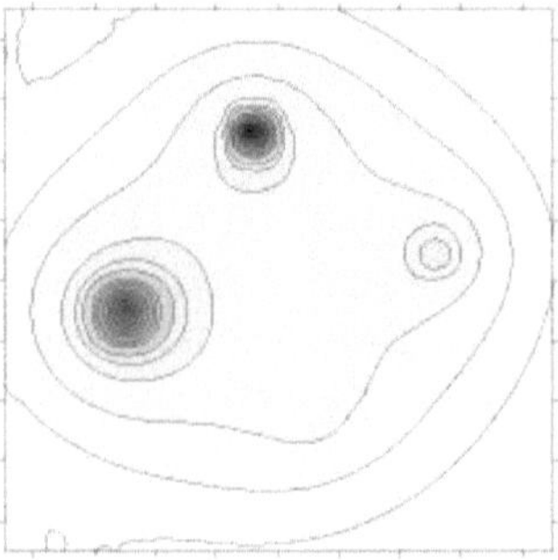

а)

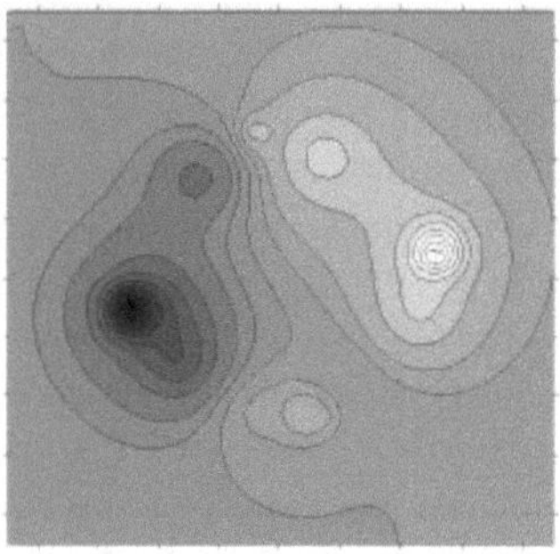

б)

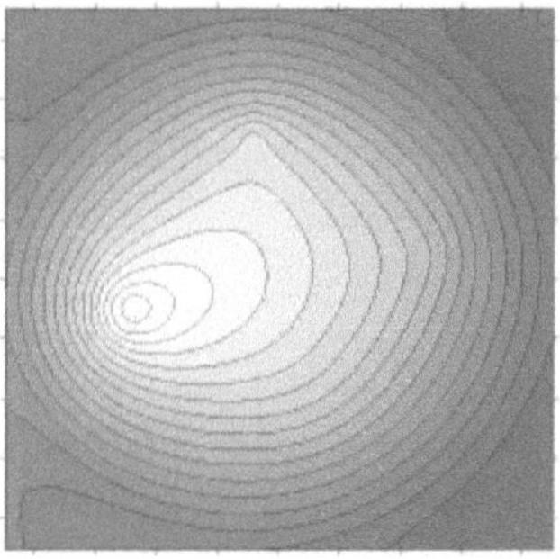

в)

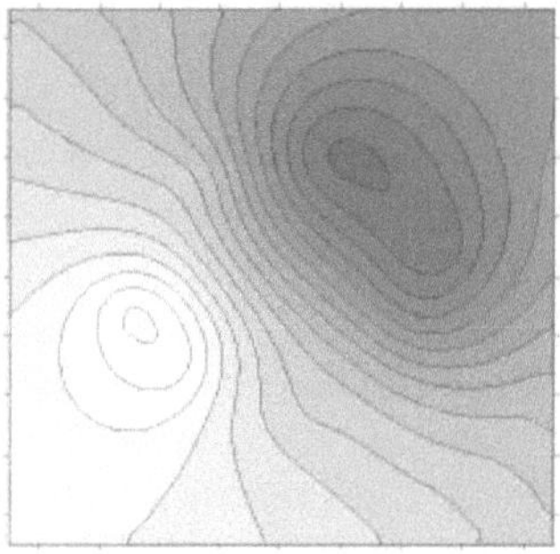

г)

Parameter maps of the divergence field divergence parameters of the atomic mass distribution IP gradient (a) and "stereochemical charges" (b), parameters of the atomic mass distribution IP potential (c) and "stereochemical charges" (d) for the host molecule $R_{\cdot h}$

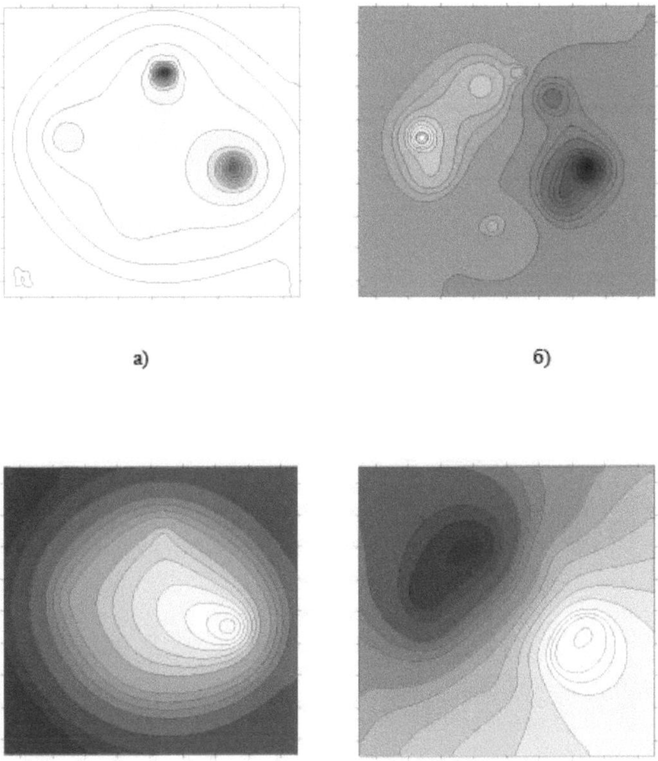

а) б) в) г)

Parameter maps of the divergence field divergence parameters of the atomic mass distribution IP gradient (a) and "stereochemical charges" (b), parameters of the atomic mass distribution IP potential (c) and "stereochemical charges" (d) for the host molecule S

·h

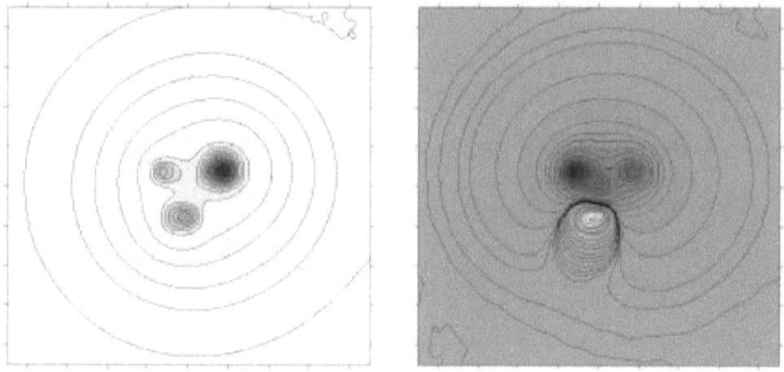

а) б)

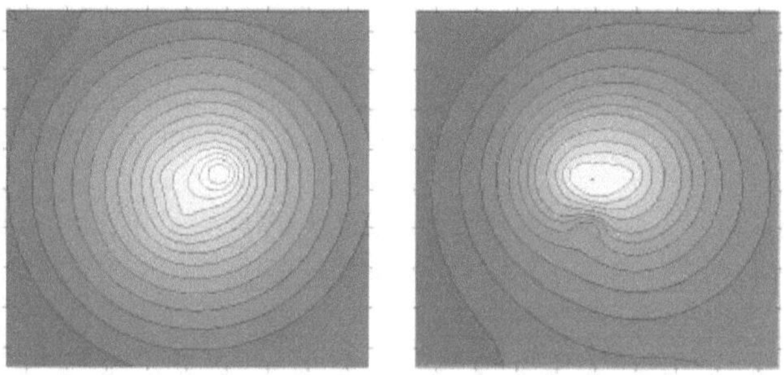

в) г)

Parameter maps of the divergence field divergence parameters of the atomic mass distribution IP gradient (a) and "stereochemical charges" (b), parameters of the atomic mass distribution IP potential (c) and "stereochemical charges" (d) for the guest molecule R .g

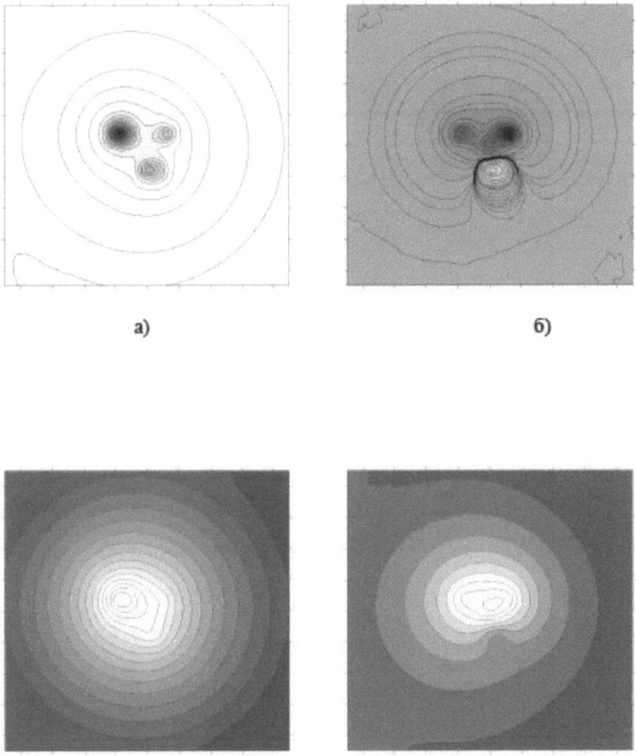

а) б) в) г)

Parameter maps of the divergence field divergence parameters of the atomic mass distribution IP gradient (a) and "stereochemical charges" (b), parameters of the atomic mass distribution IP potential (c) and "stereochemical charges" (d) for the guest molecule S $_{\cdot g}$

APPENDIX B
Steroid structures

Структуры стероидов

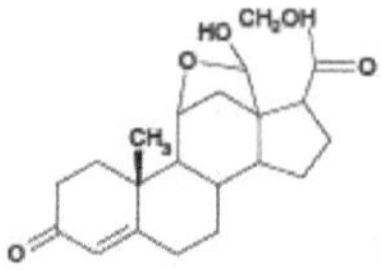
1

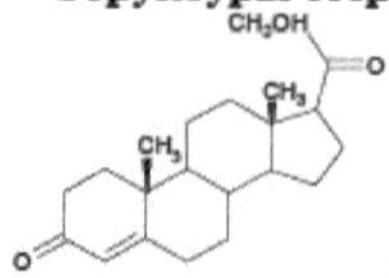
2

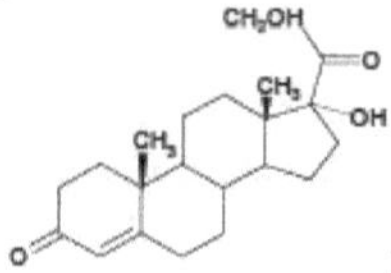
3

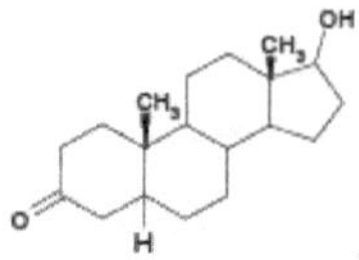
4

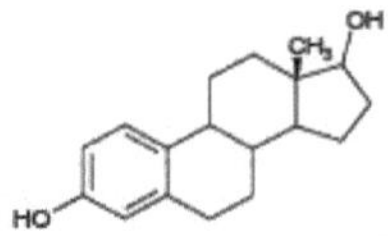
5

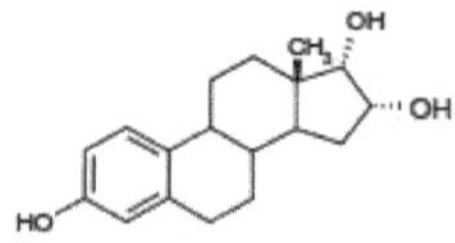
6

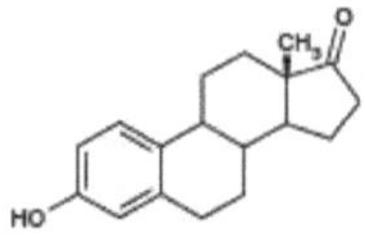
7

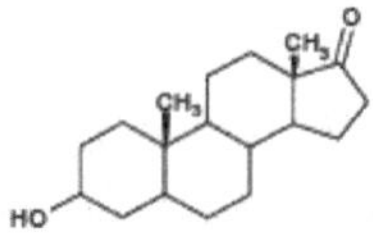
8

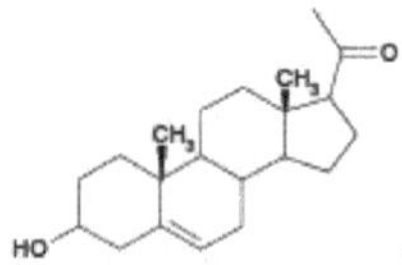
9

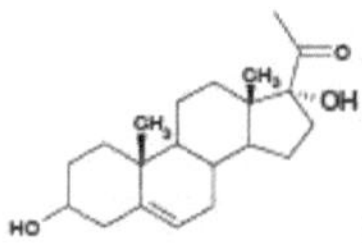
10

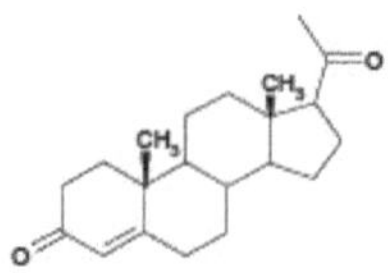
11

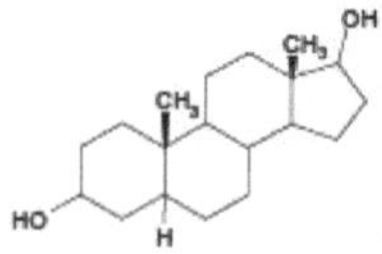
12

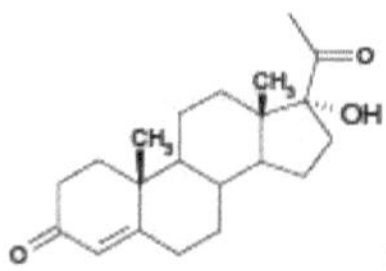
13

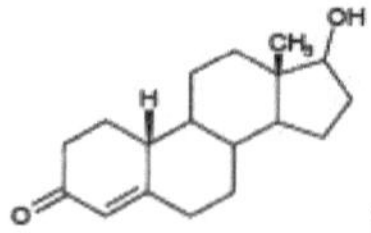
14

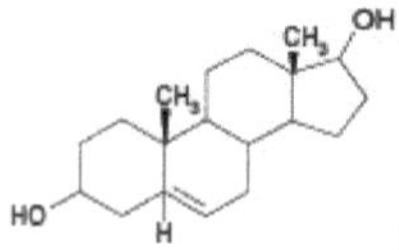
15

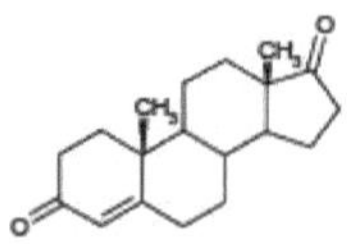
16

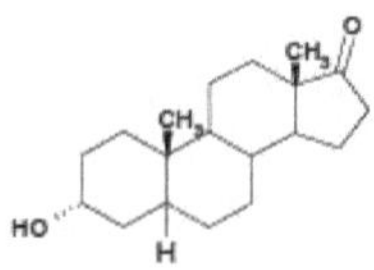
17

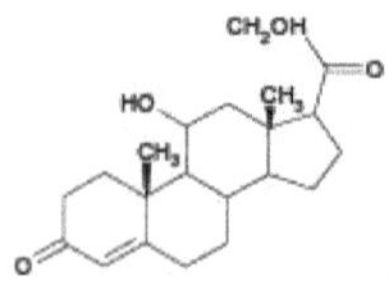
18

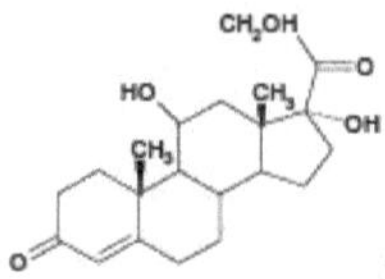
19

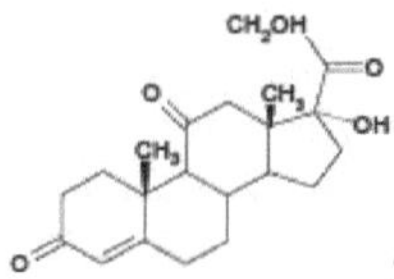
20

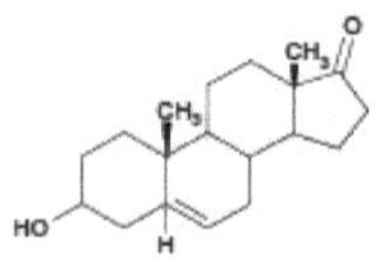
21

MIX
Papier aus verantwortungsvollen Quellen
Paper from responsible sources
FSC® C105338

Printed by Books on Demand GmbH, Norderstedt / Germany